Distance Learning
Technology and Applications

For a complete listing of the *Artech House Telecommunications Library*, turn to the back of this book.

Distance Learning Technology and Applications

Daniel Minoli

Teleport Communications Group
Stevens Institute of Technology

Artech House
Boston • London

Library of Congress Cataloging-in-Publication Data
Minoli, Daniel, 1952–
   Distance learning technology and applications / Dan Minoli.
      p.   cm.
   Includes bibliographical references and index.
   ISBN 0-89006-739-2 (alk. paper)
   1. Distance education—United States.   2. Education technology—
United States.   3. Telecommunications in education—United States.
4. Education—Computer network resources—United States.   I. Title.
LC5805.M55   1996
371.3'078—dc20
                                                                96-6143
                                                                CIP

British Library Cataloguing in Publication Data
Minoli, Dan
   Distance learning technology and applications
   1. Telecommunication in education   2. Distance education—
   Technological innovations   3. Open learning—Technological
   innovations
   I. Title
   371.3'078

   ISBN 0-89006-739-2

LC 5805 .M55 1996

Cover design by Darrell Judd

© **1996 ARTECH HOUSE, INC.**
685 Canton Street
Norwood, MA 02062

International Standard Book Number: 0-89006-739-2
Library of Congress Catalog Card Number: 96-6143

10 9 8 7 6 5 4 3 2 1

This time for Gabrielle

Contents

Preface

Interactive distance learning (IDL) is the interactive delivery of educational programming to remote sites. These remote sites can consist of a single individual or groups of individuals, such as a classroom or a corporate (training) center. With distance learning, students in one school district can take advantage of courses offered in another district, and employees can be retrained without having to leave their work location. According to observers, IDL "has exploded over the last several years in American education at all levels, colleges and universities, public and private schools." In the United States alone, the corporate training market is estimated to be a $100-billion-a-year business. More than 35 million individuals receive formal, corporate-sponsored education each year, and a fair portion of that market can be served by distance learning methods. K–12 education and postsecondary education are also very large markets in which distance learning is expected to make inroads in the next few years. K–12 is nearly a $400-billion-a-year industry, and postsecondary education is about $150 billion a year.

Education and training will acquire added importance in the next few years as the trend toward an information-based economy continues to intensify. In the 1970s and 1980s, low productivity growth and competitive weaknesses of the U.S. manufacturing industry resulted in the loss of blue-collar jobs to overseas countries. Currently, the service sector is a predominant portion of the U.S. economy. However, what is perceived to be "unimpressive performance" by the service sector is fueling a migration of certain white-collar jobs (e.g., software development) to overseas locations, which raises concern about potential parallel migration of blue-collar jobs. One way to combat such job losses is through an aggressive educational program of lifelong learning, and IDL may well be the approach to make that happen.

Supportive technologies such as high-speed networks (providing analog and/or dedicated connectivity) and broadband networks (providing digital and/or switched connectivity) will be important in making video-based IDL a commonly available service, although many existing systems also employ lower communication speeds. Digital video compression and distribution techniques are equally important.

This book has been written in recognition of the importance of distance learning and in support of a growing distance learning industry. It is targeted to technology providers, who need to understand the dynamics of IDL to be able to supply the industry with appropriate telecommunication services, as well as to technology consumers, who need to understand the basics of available telecommunication technologies and tradeoffs among alternatives.

The industry is attracting an increasing number of stakeholders, including the following six industry groups:

- Education providers (universities, corporate trainers, educational TV providers, and statewide distance learning networks) who are either implementing distance learning programs or are considering the implementation of such programs.
- Education receivers, including high school students, college students, and corporate employees, who have a diverse range of educational needs and are seeking distance learning as an alternative to traditional face-to-face educational methods.
- Network service providers, such as *interexchange carriers* (IXCs), *local exchange carriers* (LECs), *alternate access providers* (AAPs), and cable television companies. These service providers are becoming increasingly competitive and are racing to create a national infrastructure in support of the needs of the distance learning industry as well as the needs of other stakeholders.
- Providers of applications solutions, such as video conferencing, video broadcasting, and audiographics, who are exploring new business opportunities.
- Government agencies at the state and federal levels that are attempting to speed up the process of creating a *national information infrastructure* (NII) and to create a better education system.
- The Internet community, which continues to grow at a rapid pace and to expand the range of educational services.

The primary goals of this book are to provide these participants with what we hope to be an encompassing guide to the business, technical, and regulatory factors that are shaping the growth of distance learning industry and to propose specific business and application solutions that can speed the process of industry development. Specifically, this book:

- Outlines the major education problems facing society;

- Identifies the major current and potential receivers of distance learning services and outlines the factors driving their needs for distance learning;
- Defines the key communications applications of distance learning and analyzes communications requirements of distance learning providers and receivers, particularly broadband and video;
- Describes a number of major application and networking solutions that address the requirements of distance learning providers and receivers and compares the strengths and shortcomings of those solutions;
- Describes current efforts of the network service providers to build an NII in support of distance learning applications (as well as other applications) and highlights the key limitations of the current infrastructures;
- Explores the range of services that will ride on this infrastructure;
- Explores the current roles of government agencies in speeding the process of developing an NII;
- Proposes specific actions that participants in the distance learning industry may want to consider in speeding the process of creating an NII.

Several different groups in the distance learning industry can benefit from this book.

- Education providers can learn about the links they need to establish with education receivers, the applications driving their needs for those links, the potential applications and networking solutions on which they can rely in establishing those links, and the strengths and limitations of each solution.
- Corporate employers and employees who are considering implementing or participating in a distance learning program can acquire a sense of the economic investments and paybacks associated with such a move.
- Public policymakers and members of regulatory bodies at the state and federal levels can gain a better understanding of the current status of distance learning initiatives and opportunities, as well as the requirements they impose on the NII. This text provides specific examples of successful state government policies that are being implemented and that can further enhance the growth of distance learning.
- Education providers, university administrators, and *information systems* (IS) managers can gain a better understanding of the application and networking solutions available to them to meet the distance learning needs of potential clients, namely, corporations, high schools, and smaller colleges.
- Administrators and IS managers of education receivers, such as K–12 schools, can learn about the existing and emerging networking solutions that are (or will be) available to them and the strengths and limitations of each solution.

- Network service providers can learn about the needs of the distance learning market and to what extent each of the services they offer can meet those needs.

Although a number of books have been devoted to the topic of IDL, this book aims at making three new contributions:

- It provides a fairly detailed analysis not only of the needs of distance learning users but also of the roles of the other stakeholders in the distance learning industry, including the roles of cable TV providers, telephone companies, AAPs, state and local governments, and the Internet community.
- It provides a detailed analysis of the leading edge technologies associated with the distance learning industry and to what extent those technologies, including *asynchronous transfer mode* (ATM), fast packet services, and data over cable TV facilities, can meet the needs of distance learning users.
- It also addresses the current state of an NII as the emerging platform in support of multiple applications, including distance learning, and proposes specific actions to speed the process of creating an NII.

Following the introduction in Chapter 1, the demand side of the distance learning industry is discussed in Chapter 2 (K–12 schools), Chapter 3 (corporate needs), and Chapter 4 (universities). Chapter 5 describes many of the statewide initiatives underway at press time. Part II explores the supply side of the industry. Chapter 6 looks at the NII from a government point of view, and Chapter 7 from a telephone carrier perspective. Chapter 8 discusses cable TV companies, while Chapter 9 examines the Internet community in the context of IDL. The last three chapters provide real-life case studies: Chapter 10 addresses issues in deploying systems in K–12 schools; Chapter 11 examines New York University's Virtual College; and Chapter 12 discusses the battery of IDL networks provided by Wescott Communications (one such network was used by one of the authors in the delivery of a full-motion, interactive, nationwide seminar).

Acknowledgments

Mr. Minoli worked on this book while he was at DVI Communications in New York City; he is currently with Teleport Communications Group (TCG), also in New York City. TCG is the largest AAP and a major player in broadband communications, including ATM and video. Mr. Minoli would like to thank Mr. Roy D. Rosner, vice president, TCG Data and Private Lines, for his support of the final phase of this project. Mr. Minoli would also like to thank Mr. Ben Occhiogrosso, president, DVI Communications, for the technical input provided during the development phase of this project; such input was based on the practices developed in support of numerous education-industry clients.

The author would like to warmly thank Mr. Marc P. Pfeiffer, director of product marketing at Newbridge Networks, for his insightful treatment of the real-life case study in Chapter 10, which he authored.

The author also thanks Richard Vigilante, director of the Information Technologies Institute at New York University's School of Continuing Education for contributing Chapter 11, which describes the Virtual College. Mr. Vigilante developed the Virtual College Teleprogram to deliver interactive multimedia telecourses in information technology to student PCs over dialup digital telephone lines. The chapter is included to provide an example of an actual distance learning program.

In addition, the author thanks Mr. Bryan Polivka, vice president of programming for Westcott Communications, Inc., and Mr. Scott Finley, programming manager for the Westcott Healthcare Teleconference Group, for providing the case study in Chapter 12; also Mike Mooney, Dr. Linda Harrington, Jo Streit, and Fred Sylvester for their contributions. Westcott Communications operates and programs national networks for distance learning that reach subscribers in the workplace and the classroom. This material provides a case study of a business that has been a leader in distance learning since the mid-1980s.

Ms. Cecilia Fiscus assisted in researching a number of the topics treated in the book. The author greatly appreciates her unending support.

This book does not reflect any policy, position, or posture of any corporation or institution. All ideas expressed are those of the author. The writing of this book was not funded by anyone. Data pertaining to public networks were based on the open literature and were not reviewed by any carriers. Vendor products, services, and equipment are mentioned solely to document the state of the art of a given technology; the information has not been counterverified with vendors. No material contained in this book should be construed as a recommendation of any kind.

Part I
The Demand Side of the Distance
Learning Industry

According to observers, interactive distance learning (IDL) "has exploded over the last several years in American education at all levels, colleges and universities, public and private schools." The *National Educational Telecommunications Organization* (NETO) estimated recently that there are more than 110 domestic IDL program suppliers. The U.S. corporate training market alone is estimated to be a $100-billion-a-year business, and upward of 35 million individuals receive formal, employer-sponsored education each year. A fair portion of this market can be served by IDL methods. K–12 education and postsecondary education are also very large markets where IDL is expected to make inroads in the next few years. According to the U.S. National Center for Education Statistics, K–12 was a $375-billion business in the early 1990s, and postsecondary education was $150 billion; this is a total of over a half-trillion dollars a year. Some estimate the combined IDL, multimedia, and electronic training as being able to reach a potential share of 25–35% of the total education market.

Part I of this book analyzes the following three key demand drivers for IDL:

- K–12 schools (Chapter 2);
- Universities (Chapter 3);
- Corporations (Chapter 4).

Chapter 2 focuses on the interactive distance learning needs of K-12 students. Chapter 3 focuses on the interactive distance learning needs of universities. Chapter 4 is devoted to the interactive needs of corporations.

Introduction to the Distance Learning Environment

1

This chapter is an introduction to the distance learning industry. *Interactive distance learning* (IDL)[1] is technology-based education, free from the limitations of distance, place of learning, and even time synchronicity [1–8]. In the 1980s the Council of Postsecondary Accreditation and the State Higher Education Executive Officers Association completed a two-year study assessing distance learning via telecommunications. In their report they gave the following definition: "Telecommunications instruction is any course offered by a (postsecondary) educational institution, consortium of institutions, or other organization, for which credit is offered or awarded toward a certificate, diploma, or degree. The course or courses must have as the primary mode of delivery television, video cassette or disc, film, radio, computer, or other supportive devices which build upon the audio-video format." In many instances, the telecommunications course is supported by textbooks, study guides, library resources, and other study aids, and it also may involve personal interaction with faculty, tutors, or other educational personnel by telephone, two-way video, mail, or in face-to-face meetings.

IDL is an evolving paradigm of instruction and learning that attempts to overcome both the distance and time constraints found in traditional classroom learning. It is a set of technologies that can allow for a more equitable distribution of resources, as well as a more personalized learning experience. This type of educational concept offers instruction that, depending on the kind of system or technology, has most or all of the following characteristics: it is self-scheduled, self-motivated, and self-paced; there is no travel time; and it offers high retention, continuous availability, and a nonthreatening learning environment.

---

1. Sometimes this field is referred to as distance learning or telecommunications-based instruction. We use the term interactive distance learning to emphasize that the learning experience is two-way and interactive and that the source of the instruction or information may be at a distance and has to be accessed over an appropriate network.

Technology has been used in education beginning with instructional radio, film, and television in the early 1960s. Early versions of those media provided low-resolution images, since the technology of recording electronic signals on moving wire or magnetic tape was slow to develop. Solid-state electronic advances made possible a major growth of cable TV systems, satellite communications, and computation power. Increased precision of electronics in view of the development of digital circuitry and mass production of integrated circuit chips resulted in many new functions, lower prices, and better quality and reliability for communications systems.

The educational community, however, has been relatively slow in adopting the potential power of the new telecommunications systems. The use of computers as a tool for delivering education was implemented and experimented with in the late 1970s and early 1980s. Politicians and government allocated money for computer education, and educational software packages proliferated. Those packages followed, for the most part, a programmed learning paradigm that emphasized drill and practice. Boring, unimaginative software was largely to blame for the failure of the computer as a learning tool. By the end of the 1980s, when better software became available, publishers were uninterested, and teachers were dubious and afraid of the high cost of computer technology. It was only in the early 1990s that telecommunications-based education started to realize its potential, especially with the advent of high-power *personal computers* (PCs), broadband communications, and digital video. Perhaps as a climax to those investigations, in 1995 the Clinton administration set a national goal to connect all K–12 schools to the Internet by the year 2000; in California, a state with an aggressive agenda, the goal was to connect all schools by the end of 1995.

In the United States alone, there are 110,000 K–12 schools, 249,000 daycare centers, 28,600 libraries, 2,807 universities and colleges, 6,000 hospitals, 12,500 university-related research labs, 11,000 commercial labs, 726 federal research labs, 600,000 corporate offices, 20,000,000 retail establishments, and 58,000,000 privately owned homes [9]. Although the market potential for IDL supporting these constituencies is extensive, observers claim that telecommunications providers, both service and equipment vendors, were not initially interested in the distance learning market. But in the early 1990s, the market potential was recognized, and the pace of telecommunications assessment, introduction, and use in the education field has since increased significantly. Penetration has been driven by the demand pull from educational institutions as well by the supply push from telecommunications service and equipment providers. Just the telecommunication component of IDL was estimated at $2.3 billion in 1995 [10].

This book was written in recognition of the importance of distance learning and in support of a growing distance learning industry. It is targeted at technology providers, who, to supply the industry with appropriate telecommunication services, need to understand some of the dynamics. It is also targeted at technology consumers, who need to understand the basics of available telecommunication technologies and the tradeoffs among alternatives.

This chapter discusses the educational challenges that face contemporary society. It introduces the concept of distance learning and presents the key drivers of, and the inhibitors to, the growth of the distance learning industry. The reader is then introduced to the key elements of the distance learning industry, elements that are further examined throughout the book. A brief description of available technologies follows. The chapter concludes with a description of the structure and logical sequence of the remaining material in this book.

1.1 CHALLENGES FACING SOCIETY

Education and training are strategic tools that a society needs to continuously apply in order to sustain a global competitive advantage and to create a better standard of living. The U.S. education system faces a number of challenges as it enters the twenty-first century. These challenges threaten to erode the leading position of the United States in the global economy and the quality of life of its citizens. Major U.S. institutions, including corporations, high schools, universities, and government agencies, all face education-related challenges. The following is a summary of these challenges, which are explored in more detail throughout the book.

1.1.1 Corporate Challenges

U.S. corporations face a number of major challenges. One such challenge is growing competition from Japan and the European Union. Global competition will intensify in the coming years as other Southeast Asian countries join the developed countries in competing head to head with the United States. The second challenge facing corporations is the issue of how to prepare the workforce for effective participation in the skill-intensive industries, which increasingly represent the backbone of the U.S. economy. Examples of these industries include aerospace, biochemicals, and telecommunications. The United States can no longer compete in industries that are based on low-wage labor and that deliver long runs of standardized products (e.g., the textile industry). Those industries have moved to the Far East and are increasingly moving to third world countries. Skill-intensive industries are increasingly replacing traditional labor-intensive industries as the key creators of new jobs in the economy. Thus, the success of the "American corporation" will depend to a great extent on the relative skills of its workforce and to a lesser extent on the availability of natural resources and the cost of labor.

U.S. corporations face the challenge of upgrading the skills of their blue collar workforces to increase their productivity and to enhance their abilities to manage complex machines, preeminently the computer. The challenge of corporate (re)training is not limited to blue collar workers; it also applies to white collar workers, who will have to keep up with the latest developments in their fields.

1.1.2 Government Challenges

While corporations need to upgrade the skills of their employees, government agencies at the federal, state, and local levels need to deal with the problem of structural unemployment. That problem, which affects 5 million to 6 million blue collar workers in the United States, is a result of the transition of the economy from a labor-intensive economy to a skill-based economy. Structurally unemployed people lack skills and need to be retrained so they can rejoin the American workforce.

Another major challenge facing the U.S. government is how to convert the welfare system to a "workfare" system. Welfare recipients have no hope of joining the workforce unless they acquire basic skills, such as reading and writing. At the time of the writing of this book, a major item on the agenda of the 104th Congress was to develop welfare reform legislation. Any welfare bill that passes Congress and is approved by the President cannot succeed in addressing the nation's educational problems unless it deals explicitly with two interrelated problems: retraining welfare recipients and curbing illegal immigration. Illegal immigrants represent a growing strain on the U.S. public education systems, particularly in border states, such as Texas and California.

1.1.3 Challenges Facing K–12 Schools

K–12 schools in the United States are affected by a set of problems that threaten the long-term availability skills of our society and our standard of living. One of the challenges facing K–12 schools here (and in other countries as well) is *education inequity*. Rural communities are becoming increasingly isolated as the resources available to them decline. Specifically, one resource that is increasingly lacking in rural communities is *quality* teachers. As a result, such communities are becoming "information poor" compared to more affluent suburban communities. The inner cities also face the same problems encountered by rural communities. The desire for equal access to education, particularly at the K–12 level, has led to a number of lawsuits in recent years against what is considered to be inequitable state funding of schools. Throughout the early 1990s, high courts in several states have declared the current system of funding unconstitutional. Distance learning technology has been used to address those inequities in, for example, the states of Kentucky, Texas, and Montana (Texas and Montana have made major commitments of state funding for educational technology).

Another major problem facing K–12 schools is the growing shortage of science and math teachers. According to some industry forecasts, by the year 2000, the number of public school students could reach 44 million, an increase of nearly 10%. At the same time, only one million teachers will be available to fill two million positions. American students continue to compare unfavorably with students in other nations in subjects such as math and science, as measured by standardized testing. The state governors and the President have formulated new education goals targeted for fruition by the year 2000. Many education specialists, however, recognize that those goals are unattainable by traditional education programs alone. Distance learning techniques are seen as a the means of delivering education, particularly to remote schools.

A third major problem facing K–12 school is that the traditional classroom is physically isolated from other learning environments, such as universities, libraries, and museums. The traditional classroom is also isolated from practical sources of knowledge, such as corporations and *research and development* (R&D) institutions. In addition, most classrooms lack the electronic means (computers, modems, televisions) to access those environments.

Endogenous educational restructuring and reform are also expediting the introduction of distance learning technologies. Reformers advocate a shift away from instructor-centered learning to student-centered learning, and instruction is increasingly being delivered through computers. Schools are also placing more emphasis on collaborative learning, problem solving, and results sharing, all of which require access to the information resources available through communication networks.

1.1.4 Challenges Facing Universities

Universities are faced with their own set of problems. First, if they follow the traditional approach in delivering education, they must continue to incur the cost of building new campuses and maintaining existing ones. Another key challenge is the increasing cost of education. University tuition is on the rise, outpacing inflation.

A third problem facing most universities is the declining interest of some graduating high school students in obtaining a college education. This is forcing universities to formulate strategies to enable them to compete effectively with other universities to attract high school graduates. The decrease in potential population, due to natural demographics, has forced many institutions to undergo self-examination. Technology is one of the methods being brought to bear to support quality teaching. As colleges and universities become increasingly market-driven, they are targeting students not on the basis of geographical proximity but on the basis of interest in programs. Distance learning technology affords access to a sizable market of nontraditional students who are older, have family responsibilities, are employed, wish to remain in their geographic areas, and need continuing education to succeed more effectively at work [10].

A fourth problem is the declining support from the federal and state governments. For example, State University of New York administrators recently announced that if the cuts proposed by the governor were approved by the state assembly, the university would be forced to close a number of campuses throughout the state.

1.2 THE ROLE OF DISTANCE LEARNING

The problems highlighted so far represent substantial challenges to high schools, universities, government agencies at the state and federal levels, and corporations. These problems cannot be addressed or resolved without massive and coordinated efforts that involve all those institutions. And such efforts cannot succeed unless they are based on commitments of all the parties involved to a common national goal: the creation of a *technology-based, continuous, affordable, and equitable educational system.* Proponents see IDL as "classrooms without walls...class size without limits...teaching that transcends space and time so that teachers with valuable specialties and invaluable experience can reach students hundreds of miles away...and their students can reach back, to share questions and answers that make distance learning truly interactive" [11].

Distance learning has become an option because of the increasing availability of higher-speed, two-way digital telecommunication facilities. Fiber-based terrestrial networks are available in most U.S. markets, and advances in digital video compression means that a "low-end" DS1 line (operating at 1.544 Mbps) is adequate for many distance learning applications. In addition, satellite-based networks cover nearly the entire United States; although two-way connectivity is difficult to achieve directly, a return path can be secured using a dialup voice/data line. The bottom line is that the growing availability of connectivity options is positively influencing educational institutions' adoption of distance learning.

The road to the goals of an equitable, technology-rich education system requires a number of initiatives. IDL is one the technological tools that can contribute to the solution of the educational problems that society faces. IDL refers to a portfolio of application and networking solutions that can be (and are being) implemented by universities, K–12 schools, corporations, and government agencies to enable them to enhance the education and research process. IDL solutions establish telecommunications links between education receivers and various sources of knowledge over geographic areas that may span a metropolitan segment, a whole state, or the entire country. These sources of knowledge are not limited to distance learning providers, such as universities and corporate training centers; they also encompass information and processing resources, such as libraries, supercomputer centers, and museums. The required links may involve voice communications, video communications, data communications, or a combination of these media.

The quality of the distance learning programs is a function of the selected type of solution and the particular needs of the distance learners. Those needs and the solutions available to address them are the issues that will be dealt with in detail throughout this book.

The remainder of this chapter summarizes the key advantages of distance learning and provides a brief description of the key elements of the distance learning industry structure. Some applicable technologies are also discussed.

1.3 THE BENEFITS OF DISTANCE LEARNING

IDL solutions can address many of the challenges facing education and training institutions. The following examples are only a few of the benefits that can be accrued.

- IDL can reduce the isolation of rural K–12 schools and colleges, and it can address anticipated shortages of teachers in those communities by linking them with education providers.
- IDL can eliminate the electronic isolation of the traditional K–12 classroom by linking classrooms with external sources of learning, such as museums, libraries, and universities.
- IDL can enhance the quality of education by speeding the process of information transfer between education providers and education receivers.
- IDL offers the promise of instant sharing of information among members of the R&D community, speeding the research process and, consequently, the development of new products and services.
- IDL can resolve some of the financial problems facing universities by providing them with economies of scale. Through distance learning, a greater number of remote classes can be created, giving education providers the opportunity to generate more revenues per teacher or to reduce the tuition per student.
- By implementing distance learning solutions, a university can differentiate itself from other universities by being at the leading edge of technology.
- Through distance learning, government agencies can reduce the cost of retraining the structurally unemployed and welfare recipients.
- IDL enables corporations to upgrade the skills of their workforces to effectively compete in skill-intensive industries.

- IDL can reduce the training budgets of corporations by reducing costs of travel for educational purposes.

Educators and administrators who appreciate the positive value of telecommunications are investing school capital and operating funds in IDL technology; in many schools and colleges, telecommunications has become a line item in the operating budget.

1.4   BARRIERS TO THE GROWTH OF DISTANCE LEARNING

While distance learning affords education providers and receivers with a number of important benefits, a number of factors are likely to slow the potential growth of the distance learning market:

- *Union resistance.* Eager to protect the interests of their members, teachers' unions are likely to resist the growth of distance learning, fearing that such growth would result in the reduction of teaching jobs.
- *High capital cost.* The creation of distance learning systems requires an investment by education providers and receivers in the creation, maintenance, and operation of telecommunications systems. Those costs can be substantial and in some cases are beyond the means of the education providers and receivers interested in implementing them. Planners must understand that technology acquisition without budget commitments on programming (including recurring communications costs) and training, will not result in a successful IDL experience.
- *Teacher-related factors.* Three teacher-related factors represent barriers to the growth of distance learning. The first is "technophobia," a teacher's fear of dealing with new technologies. The second is computer illiteracy in a portion of the teacher population. The third is teacher's reluctance to switch from traditional methods of teaching to technology-oriented approaches, a switch that sometimes requires a significant amount of effort.

1.5   U.S. DISTANCE LEARNING NETWORKS AND ENTITIES

This section is a summary of key distance learning agencies and networks in the United States. The networks can be characterized as interstate and intrastate. This short review should give the reader a sense of the vitality of the industry.

1.5.1 Interstate Networks and Entities

This section provides a list of some of the key interstate IDL programs.

- *Agricultural Satellite Corporation* (AGSAT) is a consortium of 42 university affiliates and two government agencies. Both noncredit and credit courses are delivered via satellite to sites in 40 states. In addition, agricultural and agribusiness research information is disseminated.
- The *Annenberg/CPB Project* was established in 1981 and is based at the Corporation for Public Broadcasting. The original charter was to assist in the improvement of and access to higher education. The Project has funded high-quality video telecourses as well as new technology projects. More recently, the Project has adopted the goal of spearheading the reform of math and science education through technology.
- Oklahoma State University created the *Arts and Science Teleconferencing Service* (ASTS) in 1988 to oversee K–12 programming. ASTS is a partnership of rural school administrators, the Oklahoma Department of Education, and the College of Arts and Sciences at Oklahoma State University. ASTS provides live school programming via satellite in math, science, and language to approximately 425 schools; the ad hoc programming reaches more than 900 institutions.
- The *Black College Satellite Network* (BCSN) includes 65 school districts that participate with black colleges and universities to receive educational programming.
- *Cable in the Classroom* is a nonprofit service supported by the cable TV industry. It offers free service to all K–12 schools passed by cable facilities. The service includes noncommercial programming, support materials, and copyright-free use of those material.
- *Channel One* is a satellite-delivered daily news and information program reaching over 12,000 K–12 schools. The free 12-minute program includes two minutes of commercials.
- *Learning Links* is a national consortium of 22 agencies linked under the Central Educational Network. The service, which is available to K–12, supports online access to databases, bulletin boards, and e-mail.
- The *Massachusetts Corporation for Educational Telecommunications* (MCET) was established in the early 1980s and is funded by the state legislature. It distributes as well as produces K–12 instructional programming over a satellite network (Mass LearnPike) and over a computer network (Mass LearnNet).
- The *Mind Extension University* (ME/U) delivers educational programming nationwide in cooperation with cable TV systems. ME/U provides credit-based undergraduate and graduate courses from a number of universities.

- The *National Distance Learning Center* (NDLC) acts as a centralized electronic information source. It keeps detailed listings of teleconferences and K–12, higher education, and continuing education courses.
- The *National Telecommunications and Information Administration* (NTIA) is part of the Department of Commerce. It acts as the President's advisor on telecommunications policies, particularly as they relate to the nation's economy and technological base. NTIA funds the Public Television Facilities Program and the Telecommunications Information Infrastructure Assistance Program (TIIAP). TIIAP, which started in the early 1990s with a $26 million fund to assist educational institutions, has reviewed over 1,100 proposals for IDL programs.
- The *National Technological University* was founded in the mid-1980s and now includes 45 participating universities in the United States. Over a dozen master of science programs in engineering are provided by satellite to full-time employees and federal workers. The programming is received at over 430 sites (universities, corporate offices, and government locations), with more than 100,000 students.
- The *Public Broadcasting Service* (PBS) funds and distributes televised instructional material. It offers telecourses to about 2,000 colleges and universities, in cooperation with local public television stations, through its Adult Learning Service and the Adult Learning Satellite Service. PBS also offers the Elementary/Secondary Service, aimed at K–12 schools.
- The *Satellite Educational Resources Consortium* (SERC) involves state public broadcasting entities and departments of education in 25 states. It also develops instructional and professional programming.
- The *Satellite Telecommunications Educational Programming* (STEP) delivers live, interactive education over satellite links.
- The *Star School Program* was authorized by Congress in 1988 and is overseen by the Department of Education. The program disseminates awards, grants, and support for a number of IDL networks, including Iowa's statewide system, SERC, BCSN, and STEP. Other channels include the Technical Education Research Centers, and the Telecommunications Education for Advances in Mathematics and Science (providing nationwide math programming for students in grades 4, 5, and 6).
- The *United States Distance Learning Association* (USDLA) promotes the development and application of IDL using satellite, video, audiographic, and multimedia technology. It spans K–12, higher education, continuing education, and corporate training.
- The *Western Cooperative for Educational Telecommunications* includes more than 150 corporations, universities, colleges, schools, and public agencies in 19 states. Its charter is to make information and expertise in telecommunications more easily available to its membership.

1.5.2 Intrastate Networks and Entities

There are many state-level IDL initiatives. States are considered to be the fulcrum for planning in education: Most of the significant education policy emanates from the state capital, either from the department of education for K–12 schools or from the board of regents, higher education commission, or state university system [10]. This state-centric paradigm also applies to IDL. Chapter 5 covers this topic at length.

It should be noted that in the past couple of years only a few states have increased their funding of IDL. Most states continue to be financially disadvantaged, particularly in view of the 1995 budgetary reforms initiated in Congress. Nonetheless, in most states legislators, administrators, and regulators are collaborating to identify new ways to fund applications through matching programs, incentive regulation, and taxes to generate funds. For example, Missouri's VIDEO program relies on a videotape rental tax. Other states (e.g., Georgia and Michigan) have created new telecommunications rate regulations that are favorable to IDL.

1.6 STRUCTURE OF THE DISTANCE LEARNING INDUSTRY

The benefits of distance learning are driving the growth of a distance learning industry made up of providers of distance learning application and networking solutions and buyers of distance learning networking solutions, including both distance learning providers and receivers.

1.7 DISTANCE LEARNING APPLICATION SOLUTIONS

Many distance learning communication solutions are on the market today to satisfy the needs of distance learning providers and receivers. These application solutions can be divided into two categories: video solutions and data solutions (see Figure 1.1).

1.7.1 Video Solutions

Video solutions can be further classified into the following categories, in increasing order of complexity and sophistication [12]:

- One-way video/one-way audio;
- One-way video/two-way audio;
- Two-way video/two-way audio;
- N-way video/N-way audio with *continuous presence*.

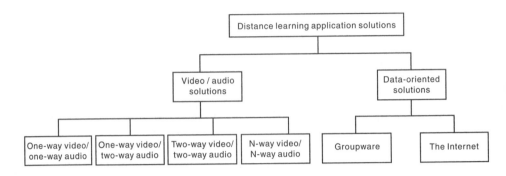

Figure 1.1 Taxonomy of distance learning technologies.

It should be noted that, despite the drive toward digital technology and fiberoptics, a major portion of distance learning programming is still carried by analog satellite links (although the use of compressed digital video is compelling), microwave links (analog links for TV distribution), over-the-air and cable analog TV, and *instructional television fixed service* (ITFS). Once such recognition is made, one can then focus on an assessment of the advantages of a migration to digital.

One-Way Video/One-Way Audio Solutions

An example of the one-way video/audio solution is educational TV programming. K–12 schools can access these TV programs by subscribing to basic cable TV services or obtain them over an ITFS network. For example, students and teachers can access educational programming about computers, communication, multimedia, software, and related matters through the Jones Computer Program. They can also watch cultural lectures, debates, and symposia through the Horizons Cable Network. In addition, they can have general education and enrichment programming through the Learning Channel. Students can also monitor world events on *Cable News Network* (CNN), which provides news coverage 24 hours a day.

One-Way Video/Two-Way Audio Solutions

These video broadcasting systems are predominantly satellite-based and enable distance learning providers to establish a one-way video/two-way audio with the recipients. While the video program originates at one site, the distance learning receivers can be located at multiple sites. Transmission is either analog or digital. Traditional analog TV requires 6 MHz; digital transmission rates range from 3 to 6 Mbps. Video broadcasting systems can be classified into two categories:

- *Private-business TV-based programming* (e.g., specialized training, seminars). These systems are used by corporations internally, for example, to provide corporate training.
- *Shared institutional distance learning systems.* These systems are shared by multiple distance learning receivers (e.g., universities, corporations) and are managed by a single distance learning provider. Examples of satellite-based distance leaning systems include BCSN, the National University Teleconferencing Network, the National Technological University, and Wescott Communications.

Two-Way Video/Two-Way Audio Systems

Two-way video/two-way audio systems provide participants with simultaneous interactive image and voice communications, enabling individuals in different locations to communicate with each other as if they were in the same room. Users may be students communicating with their teachers over a long distance or executives communicating with their employees. Videoconferencing systems involve the transmission of two-way video and two-way audio from one location to multiple locations. To reduce the amount of bandwidth required for transmission, digitization and compression of audio and video signals are implemented. One company with a full-motion two-way video training network is Bell Communications Research; an example of a compressed-video two-way video training network is Teleport Communications Group.

Video/videoconferencing systems can be classified into four categories:

- *Custom or site-built rooms.* These permanent facilities, including classrooms, are equipped with fixed systems to support video communication. Equipment is often built into walls or tabletops. The number of participants accommodated by this design is usually a function of room size and layout. This system represents the high range of video/videoconferencing systems and is appropriate for corporations that have made a major commitment to videoconferencing. High-end systems can support executive-level videoconferencing usage, and complex applications that require significant peripheral equipment. These rooms are used by both educational institutions and corporations, although the venue, the design, and the audio/visual systems tend to be different. Education institutions particularly need the following: simple control systems; monitors and projectors ranging in size from 27 inches to 53 inches (or more); three or more cameras per system (one for the student, one for the instructor, and one for graphics capture); a high-quality audio system, including quality microphones and speakers; and direct computer peripheral input to the audio/visual system. Figure 1.2 depicts the layout of a typical classroom; Figure 1.3 depicts a system-level view of an IDL site.

- *Standalone units.* These systems usually are stationary, cabinet-mounted units that contain the major equipment required for video/videoconferencing. Standalone systems accommodate the same types of applications

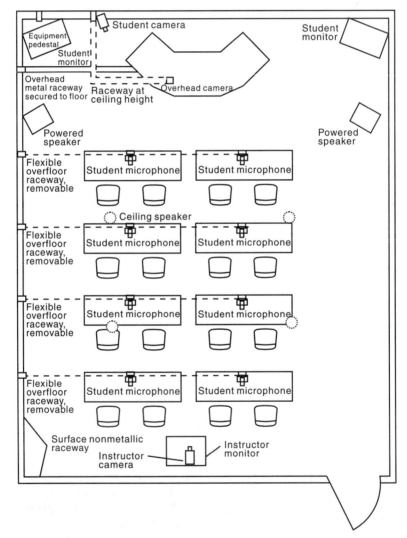

Figure 1.2 Typical distance learning classroom.

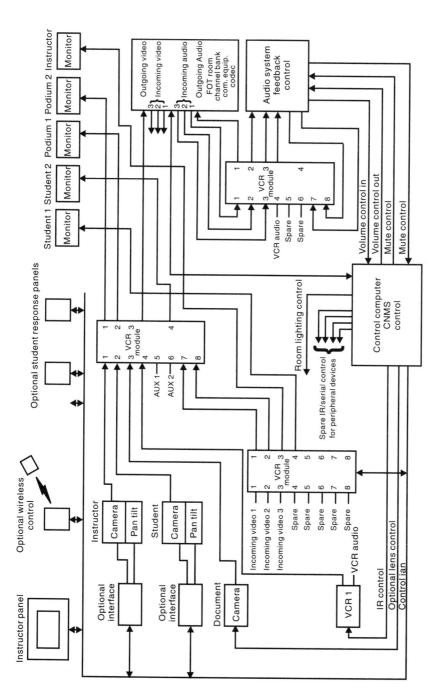

Figure 1.3 System-level view of an IDL teaching site (*courtesy of Sony*).

and the same participants as custom-built rooms. Standalone units, however, offer a less costly and less permanent solution. These systems can be used by both educational institutions and corporations.

- *Rollabouts.* These self-contained systems are configured on carts with staging rooms for storing peripherals. These systems, which represent the midrange of video/videoconferencing systems, have two advantages: (1) They are convenient because of their transportability and flexibility, and (2) they are more affordable than high-end systems. These advantages make rollabouts appropriate for companies with limited space, moderate budget and video/videoconferencing needs which may be dispersed throughout a specific location. These systems can be used by both educational institutions and corporations.

- *Desktop systems.* These PC-based systems are intended for use at employees' desks or in small meeting areas. They currently represent the low end of videoconferencing systems, making them the most affordable. PC-based video/videoconferencing systems can be created by connecting PCs to local area networks (LANs), which are extended to *wide area networks* (WANs) through services such as switched digital services. PCs can also be connected directly to WANs.

Video communication systems have the following four core components.

- *Video monitor.* Although classrooms tend to have multiple monitors, many (corporate) video/videoconferencing systems rely on one monitor to display both the incoming and the outgoing video images. This is accomplished through windowing, whereby a small window displaying one image (usually outgoing video) is superimposed on the main window (usually displaying incoming video). Alternatively, dual windows can display incoming and outgoing video or incoming video and graphics simultaneously. The number and setup of monitors depends on the application, the room characteristics, and the type of videoconferencing installation (e.g., desktop, rollabout, custom-built). Large monitors tend to be more effective.

- *Camera(s).* The number and the arrangement of cameras depends largely on the number of conference participants, students, or teachers; the application requirements; the room characteristics; and the type of videoconferencing installation (e.g., desktop, rollabout, custom-built). Cameras range from low-end systems, to *one-charged-couple-device* (1-CCD) systems with 470 lines of horizontal resolution, to 3-CCD cameras with 750 lines of resolution. Cameras can be positioned in one of three ways: with a joystick or control panel buttons, automatically to focus on the participant who is speaking, or manually.

- *Audio system.* Audio quality is as important as picture resolution. Audio systems, including speakers and microphones, must provide clear, simultaneous two-way communication. One desirable feature of an audio system is *automatic echo cancellation*, which adjusts the incoming and outgoing audio signals according to the acoustics of a particular room.
- *Coder/decoder* (codec). A codec performs two functions: video digitization and compression. Digitization is the ability of the codec to convert analog video and audio signals into a digital bit stream for transmission over digital lines. When receiving the coded signal, the remote codec reconverts the digital signal back into analog video and voice signals, which are then sent to the video monitor and speakers. The codec also compresses the signal to provide efficient and cost-effective transmission. Typically the compression ratio is 100:1. The level of picture resolution provided by the codec depends on the sophistication of the digitization and compression algorithms. Video compression algorithms are based on the *International Telecommunication Union–Telecommunication* (ITU-T) *or International Organization for Standardization* (ISO, after the French) standards. ITU-T's standard H.261 covers transmission of video at rates ranging from 56/64 Kbps to 2.048 Mbps. ISO's Motion Picture Expert Group 1 (MPEG-1) supports entertainment-level video at 1.544 Mbps; MPEG-2 supports even better quality at 6 Mbps. The *Motion Joint Photographic Expert Group* (M-JPEG) supports quality video at 10–20 Mbps [9].

Video compression algorithms are critical to the viability of distance learning and multimedia, both in terms of storage as well as communication. Multimedia and video objects are large: Digitized speech, music, still images, and, even more so, full-motion, full-quality video all require large amounts of storage and communication bandwidth. For example, uncompressed digital TV video requires 140–270 Mbps. It follows that compression schemes are the only hope for widespread deployment of digital video. Fortunately, video signals contain a substantial amount of intrinsic redundancy, so that compression can be undertaken prior to storage or transmission. Compression schemes that are able to reduce the raw bits by a factor of 100 or 200 are usually employed. There have been claims of compression ratios of 1:2,500 through the use of *fractal* methods.

Users of videoconferencing systems can also employ the following additional optional components:

- *Control panel.* The control panel acts as the interface between the participants and system equipment, facilitating management of a video communication session. The control panel incorporates the following functions:

call initiation, camera positioning, volume control, and control of peripheral equipment.

- *Inverse multiplexers* (also called *Imux*). Current low-end H.261 codecs are capable only of combining two individual 56/64-Kbps circuits into a single 112/128-Kbps transmission stream. To achieve transmission in excess of 112/128 Kbps, an inverse multiplexer is required. Current inverse multiplexers allow 2 to 48 individual switched 56/64-Kbps circuits to be combined into larger bandwidth streams.
- *Peripheral devices.* Videoconferencing systems can incorporate the following devices: PCs, fax machines, slide projectors, VCRs, telephones, and document cameras to display transparencies or photographs during a video conference.
- *Video bridges.* These devices, located at the customer location or in the network, allow the support of multipoint conferencing. This technology also supports continuous presence, which is discussed in the next subsection.

Audio/visual distance learning systems are not inexpensive and may typically require a system integrator to assemble and test. Tables 1.1 and 1.2 provide a sense of the expense per teaching site (as of press time); although these figures will change over time, the relative magnitude will remain the same.

Table 1.1
Representative Costs for a Distance Learning Site

Sony Electronics, Inc., Distance Learning Systems Price List		
Model	*Description*	*List Price*
DLSYS101PAC	Sony distance learning classroom system, including dual 53-inch display, three single CCD camera setup including pan/tilt control, audio package including auto mixer with 8 mics and room control system including echo cancelation. Does not include CODEC.	$46,840
DLSYS102PAC	Sony distance learning classroom system including dual 32-inch display, three single CCD camera setup including pan/tilt control audio package including auto mixer with 8 mics and room control system including echo cancelation. Does not include CODEC.	$37,467

Sony Electronics, Inc., Distance Learning Systems Price List

Model	Description	List Price
DLSYS103PAC	Sony distance learning classroom system including dual 32-inch display, three single CCD camera setup including pan/tilt control audio package including auto mixer with 8 mics and room control system including echo cancelation. Does not include CODEC.	$37,277
DLSYS104PAC	Sony distance learning classroom system including dual 27-inch display, 32-inch instructor display, three single CCD camera setup including pan/tilt control, audio package including mixers and 8 mics and room control system including echo cancellation. Does not include CODEC.	$37,477

Table 1.2
Representative Component-Level Prices for a Distance Learning Site

Item No.	Qty.	Mfgr.	Model	Description	Unit Price	Total
Typical Teleclassroom Equipment						
Instructor Camera						
1	1	Parkervision	ICS-2000-S1A	Camera system		$7,495.00
Student Camera						
2	1	Parkervision	SCS-2000-S1A	Camera system		$4,995.00
Graphics Camera						
3	1	Sony	VID-P100	Copy stand w/lights, lens, CCD camera		$3,500.00
Video Program Sources						
4	1	Sony	SVO-1610	VHS VCR		$765.00
5	1	Sony	MDP-1700AR	Laser disk		$940.00
Student Monitors						
6	4	Sony	KV-27S10	27" monitor		$3,000.00
Instructor Monitors						
7	4	Sony	KV-27S10	27" monitor		$3,000.00
8	1	Sony	PVM-95	9" high-res monitor w/A-B		$395.00

Table 1.2 (continued)

Item No.	Qty.	Mfgr.	Model	Description	Unit Price	Total
Student Microphones						
9	8	Shure	AMS22	Microphone		$1,760.00
Audio Group						
10		Shure	AMS-8000	8-input automatic mixer		$2,850.00
11	1	Sabine	FBX-900	Feedback exterminator		$450.00
12	1	Rane	FPM-44	4-input mic/line mixer		$399.00
13	1	Rane	FHB-1	19" rack kit		$39.00
14	1	Rane	RS-1	Mixer amp		$18.00
15	1	PEAVEY	MA 35	Speaker		$750.00
16	1	JBL	Control 5	Speaker mount		$220.00
17	1	JBL	MTC51			$69.00
	1					
Signal Processing						
18	1	Creston	VIDMUX	Audio/video switcher		$2,587.00
19	1	Horita	BSG-50	Sync generator		$289.00
20	1	LIETCH	FR-681	Video DA frame		$750.00
21	3	LIETCH	UDA-680	Video DA card		$450.00
Control System						
22	1	Creston	CTP-1500	LCD touch panel controller		$1,500.00
23	1	Creston	CNMS	Classroom control		$1,950.00
24	1	Creston	CNPWS-100	Power supply		$525.00
25	1	Creston	CNVP-3	Volume control		$525.00
26	1	Creston	CNPC-1A	Power controller		$188.00
27	1	Creston	CNTBLOCK	Terminal block		$128.00
28	1	Creston	CNIR-6	Control card		$525.00
29	1	Creston	VIDMUX	Router		$3,075.00
30	1	Creston	CNSP-112	Cable		$75.00
31	1	Creston	CNSP-XX	Cable		$250.00
32	1	Creston	CUSTOM	System programming		$4,000.00
Equipment Cabinet						
33	1	Stantron	VCSF2004325	Rack #1		$750.00
34	1	Stantron	VCSF2002825	Rack #2		$750.00
35	1	Stantron	As required	Rack accessories		$400.00
36	2	Mid Atlantic	RSH-XXX	Rack shelves		$300.00
Miscellaneous Hardware						
36	1	Sony SID	Video cables			
40			connectors			

Continuous Presence

Continuous presence is the high-end approach to distance learning. In this environment, the instructor has control of multiple remote and local cameras, which can be panned and zoomed. The delivered signal includes voice, video, and data. Individual students may have a PC on their desks so they can directly receive the instruction, including individualized information, tests, assignments, and so on. Each classroom also has a projection screen where a "quad-split" picture of the instructor and up to three other classrooms is visible, to give a sense of being "present" (hence, the term *continuous presence*).

For example, a student in classroom A may want to ask a question or do a joint project or presentation with a student in classroom B. The projection screen in classroom C would show classroom A, classroom B, the instructor, and, say, another classroom or some display such as a slides window. The instructor can inject live feeds into one quad-split window (say, a news broadcast) or some predeveloped video clip on videodisk or *compact disc–read only memory* (CD-ROM). The instructor can also inject a slide generated from a video server at the originating site or viewgraphs (computer-based or from a manual overhead projector). *Subconferencing* also can be supported, in which a private conversation among a few parties can take place. The best example of continuous presence is a high-end news program in which there are several guests at several remote sites (perhaps several continents) with the interviewer at the center of the action (the comparison breaks down, however, unless the reader positions himself or herself at the studio, where interactivity can be supported).

Continuous presence is an approach that will see more widespread use in the future. One prototypical network has emerged in the state of North Carolina to serve K–12 students; 3,000 classrooms are expected to be connected by 1996 (see Figure 1.4).

1.7.2 Data Application Solutions

Data solutions can be classified into two categories:

- *Groupware*, which allows remote training through PCs connected to the training site by a switched or permanent network, usually employing *computer-supported cooperative work* (CSCW) software (e.g., New York University's Virtual College);
- *The Internet*, which supplies ad hoc access to information and databases.

Groupware

Groupware refers to software that can support at least one of the following applications: electronic messaging, data conferencing, or messaging gateways.

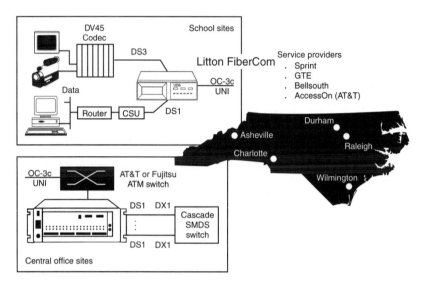

Figure 1.4 North Carolina Information Highway.

Groupware solutions link distance learning providers and receivers that are separated by a WAN. An increasing number of vendors offer groupware. Providers of groupware solutions extend beyond software providers (e.g., Lotus) to encompass network providers (e.g., MCI and AT&T). The products offered by groupware vendors can be classified into three categories:

- *Low-end groupware*, which supports a single application, such as computer conferencing, project management, workflow management, and time management;
- *Groupware that supports a single local area or workgroup* but that may support a number of functions;
- *High-end software*, which supports multiple functions over wide-area enterprise networks.

The Internet

The Internet is an international network that supports over 34,000 computer networks and over 5 million computers worldwide. Estimates of the number of Internet users vary from 7 million to 12 million; half the Internet users reside in the United States. The Internet has been growing rapidly in the last few years. Several statistics reflect that fact: The number of subscribers has been growing at 100% per year for the last 3 years; the number of universities and research laboratories using the Internet has increased from 200 in 1988 to more than 10,000 in 1993; the number of host computers grew from 1 million in 1992, to 2 million in 1994, to 5 million in 1995; and the number of packets carried on the Internet has increased from millions per day to billions per day. Through the Internet a user can receive a number of services, including logon services, e-mail, file transfer, host-to-host communications, and directory services.

Recently, the Internet has "burst into the collective consciousness of the American business community, triggering exponential growth" [13]. These changes continue to be fueled by a number of technological and related trends:

- An upsurge in the sales of home and small-business PCs and modems;
- The availability of low-cost, easier-to-use Web-browser software;
- The rapid expansion of local Internet service providers;
- The provision of full Internet access, including Web service, by commercial online services, such as Prodigy, CompuServe, America Online, and Microsoft's Windows 95;
- The entry of a large number of recent college graduates with Internet experience into the workforce.

Many areas of business, education, and consumer life are likely to be affected by further developments of the Internet.

1.8 DISTANCE LEARNING NETWORKING SOLUTIONS

Several network solutions exist to support the application solutions described in Section 1.7. It should be noted that communication systems are specific to topological segments of the network, for example, the access, the *local-access transport area* (LATA) backbone, and the national backbone, in the sense that different technologies may be employed in each segment.

Topologically, communication occurs at the following levels:

- A local- or campus-area network; that is, within a building or within a campus;
- Public network access; that is, between the user's building and the carrier's point of presence, for example, a local central office;
- A *metropolitan area network* (MAN); that is, at the community level;
- A WAN; that is, regionally or nationwide;
- A global area network; that is, internationally.

As implied, each segment of a communication link often utilizes distinct technologies. In addition, each segment is subject to different cost, performance, and availability considerations. By and large, LAN issues relate to initial wiring and the deployment of appropriate servers. Bandwidth generally is not an issue, since one can obtain from 10 to 100 Mbps using traditional methods.

Also, cost tends not to be an issue, since after the initial install (say, $400–$600 per connected workstation), there are no recurring charges. Access to public networks is currently a major area of discussion as it relates to carriers' efforts to improve it (access is what has been called metaphorically "the ramps to the information superhighway"). The factors of this discussion are: (1) the number of access lines in a network, which tracks with the number of residences and buildings requiring connectivity; and (2) the throughput of the access medium. Currently it is difficult to obtain more than 128 Kbps for a residence or more than 1.544 Mbps for a business. New applications, such as distance learning, require more bandwidth. The media to support access include twisted-pair copper lines to the domicile[2] (the staple of today's network), coaxial cable (e.g., cable TV companies), end-to-end fiber optic links, and hybrid fiber-coax systems (e.g., newer cable TV distribution systems). Existing technologies have limited throughput but are relatively inexpensive (e.g., 128 Kbps and $600 to install per location); newer technologies support much higher throughput but are more expensive, particularly in view of the number of businesses and residences that would have to be upgraded (e.g., 2–10 Gbps on fiber, with an install cost of $2,500–$4,000 per location). Technological variations are now being sought to increase bandwidth in a cost-effective manner. For example, there are ways of increasing the throughput of a twisted-pair copper line and ways of better utilizing a fiber (e.g., using wavelength division multiplexing). The issues related to MANs and WANs center on bandwidth availability, cost of the connection, and effective support of multiple locations.

From a target-market perspective, communication solutions can be classified into the following two major groups (see Figure 1.5): networking solutions that can support large and medium-sized businesses and networking solutions that support local communities.

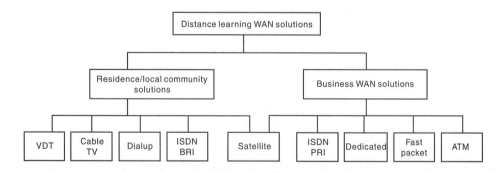

Figure 1.5   Services that can be employed to support distance learning.

---

2. The carrier may use a fiber optic link for a portion of the link. However, because this fiber is shared among hundreds of users, the amount of bandwidth to a specific user remains low.

1.8.1 Networking Solutions for Large and Medium-Sized Businesses

These networking solutions can be classified into the following:

- Dedicated services, ranging in speed from fractional *digital signal* (DS) 1 (multiples of 64 Kbps, up to 1.544 Mbps) to a *synchronous optical network* (SONET)[3] (multiples of 155 Mbps) [14,15];
- Circuit switched services, which include switched DS1 (1.544 Mbps), and switched DS3 (45 Mbps) services [14];
- *Asynchronous transfer mode* (ATM)-based solutions, which include ATM cell relay service, as well as support of *switched multimegabit data service* (SMDS) and *frame relay service* (FRS) over the same integrated platform [16,17];
- Discrete fastpacket services, which include SMDS, FRS, and *native-mode LAN interconnection service* (NMLIS) [18,19];
- *Integrated services digital networks* (ISDN), which range in speed from 64 Kbps to 1.544 Mbps (DS1) [14,15,20].

Following are brief descriptions of these networking solutions.

Digital Private Lines (Fractional DS1, DS1, DS3, SONET)

Dedicated "private line" digital services provide transparent bandwidth at the specified speed ($n \times 64$ Kbps for FDS1, 1.544 Mbps for DS1, and 44.736 Mbps for DS3) and are suitable for point-to-point interconnection of low-burstiness (i.e., steady) traffic. DS3 facilities are increasingly available in many parts of the country, although they are still relatively expensive. In addition, a number of carriers have started to offer a fractional DS3 service that allows the user to specify the desired number multiple of DS1s. These dedicated facilities can support voice, video, and data.

Frame Relay Service

FRS is a data communication service that became available in the early 1990s. It supports medium-speed connections between user equipment (routers and private switches in particular) and between user equipment and carriers' frame relay network equipment (i.e., public switches). The frame relay protocol supports data transmission over a "connection-oriented" path;[4] it enables the

---

3. Outside the United States, the digital hierarchy embodied in SONET is known as *synchronous digital hierarchy* (SDH).

4. A connection-oriented path is one that is set up at the initiation of the transaction in such a manner that all protocol data units take the same route through the network.

transmission of variable-length data units, typically up to 4,096 octets, over an assigned virtual connection. FRS provides interconnection among n (backbone) sites by requiring only that each site be connected to the "network cloud" via an access line (compare this with the $n(n-1)/2$ end-to-end lines required with dedicated services). The cloud consists of switching nodes interconnected by trunks used to carry traffic aggregated from many users. In a public frame relay network, the switches and the trunks are put in place by a carrier for use by many corporations. Carrier networks based on frame relay provide communications at up to 1.544 Mbps (in the United States), shared backbone bandwidth on demand, and multiple user sessions over a single access line. The throughput is higher than that available for traditional packet switching, making the service attractive for applications that involve LAN-based communication, including image transfer, but not voice or video. In a private frame relay network, the switches and trunks are put in place (typically) by the corporate communications department of the company in question. In either case, the service can provide *permanent virtual connections* (PVCs) and now also *switched virtual connections* (SVCs).

Switched Multimegabit Data Service

SMDS is a high-performance, public, connectionless[5] service developed by Bellcore (Bell Communications Research) in the late 1980s. Access to the network is based on a subset of the *Institute of Electrical and Electronics Engineers* (IEEE) 802.6 MAN standards. From an interface point of view, access to SMDS is established through a *subscriber network interface* (SNI), which may provide either a DS3 access path (with multiple access classes) or a DS1 access path. For a DS3 access path, single or multiple *customer premise equipment* (CPE) access arrangements are available. The access protocol, *SMDS interface protocol* (SIP), is based on *distributed queue dual bus* (DQDB) connectionless protocol. This protocol operates at the *media access control* (MAC) level of the data link layer (layer 2) of the *Open System Interconnection Reference Model* (OSIRM). As a result, SMDS can be supported by various higher-layer protocols, such as *transmission control protocol/Internet protocol* (TCP/IP).

SMDS supports five service-access classes. One of these access speeds can be supported by a DS1 access line, while the other speeds (4 Mbps, 10 Mbps, 16 Mbps, 25 Mbps, 34 Mbps) can be obtained by using a portion of a DS3 line. SMDS also operates at 64 Kbps. SMDS supports *data-oriented* (not video) distance learning application, such as groupware and the Internet, and is applicable to corporate distance learning applications, such as corporate training.

---

5. In a connectionless environment, each protocol data unit traverses the network independent of the previous unit.

ATM Cell Relay Service

ATM refers to a high-bandwidth, low-delay switching and multiplexing technology now becoming available for both public and private networks. It supports voice, video, and data. While ATM in the strict sense is simply a (data link layer) protocol, the more encompassing ATM principles and ATM-based platforms form the foundation for a variety of high-speed digital communication services aimed at corporate users for high-speed data, LAN interconnection, imaging, video, and multimedia applications. *Cell relay service* (CRS) is one of the key services enabled by ATM. CRS can be utilized for networks that use completely private communication facilities, completely public communication facilities, or that are hybrid. A variety of vendors now have ATM products on the market. A number of carriers (e.g., Teleport Communications Group) either already provide services or are poised to do so in the immediate future. ATM supports both switched (SVC) and nonswitched (PVC) connections. ATM supports services requiring both circuit-mode and packet-mode information-transfer capabilities. CRS supports both data-oriented distance learning applications such as groupware and the Internet, as well as video-oriented distance learning applications. ATM employs SONET facilities as the transport medium. These fiber-based facilities support data rates that are multiples of 155 Mbps (up to 10 Gbps).

1.8.2 Networking Solutions Supporting Local Communities

A number of networking solutions either exist on the market today or are emerging to support the data and video networking needs of local communities. Effectively, these solutions are access technologies (i.e., the metaphoric ramps onto the information superhighway). They encompass networking solutions currently offered by the local telephone companies, such as dialup and ISDN *basic rate interface* (BRI). They also encompass solutions offered by cable TV companies. In addition to these existing solutions, a number of broadband solutions are being introduced or tested by the *local exchange carriers* (LECs), including *video dialtone* (VDT) solutions. Following are brief descriptions of each solution, with more details provided throughout the book.

Asymmetric Digital Subscriber Line

Asymmetric digital subscriber line (ADSL) is an emerging high-performance copper-based technology. It is intended to provide more bandwidth out of the existing loop apparatus (i.e., the copper-based plant). ADSL is currently being evaluated or trialed by some RBOCs. Effectively, it is a repeaterless DS1 technology, but the bandwidth is asymmetric. In the forward channel (network-to-user), ADSL supports DS1 rates (1.544 Mbps) at 18 kft from the *central office*

(CO) and T2 rates (6 Mbps) at 12 kft from the CO. It also has *one bidirectional 160-Kbps ISDN basic rate channel* containing two "B" channels, a "D" channel, and associated operations channels. Hence, the reverse channel (user-to-network) is much more limited. The key targets of ADSL-based services are residential customers, local community colleges, and K–12 schools. Consequently, ADSL is applicable to the distance learning access applications that involve those local communities. ADSL can be applied to both video- and data-oriented distance learning applications, particularly Internet access, given its capabilities. It also can be used to deliver VDT services. The future of ADSL may be limited, however, because many carriers prefer to channel the nontrivial per-loop investment required to provide it toward the constuction of a broadband fiber-based access loop.

The narrowband and wideband solutions described here provide substitutes to each other, each with a set of strengths and weaknesses. Some of these solutions can also complement each other, providing users with hybrid solutions. Examples of complementary networking solutions include FRS/ISDN, cable TV/traditional telephony, and ATM/SMDS.

Hybrid Fiber Coax

A discussion of ADSL necessitates a discussion of *hybrid fiber/coax* (HFC), although ADSL technology is being pursued (if at all) by the traditional LECs, while HFC is being pursued by the cable TV companies as well as by the LECs. At this time, most of the cable TV industry is working on upgrading the traditional, all-coax network to fiber/coax hybrid systems. In many cable TV networks today, fiber is being used at least in the trunk portions of the network; as time goes by, fiber will make even more inroads. Ultimately there is a desire—and a need—to provide fiber connectivity all the way to a residence or an institution. This, however, is still fairly expensive. The telephone carriers, namely, LECs and *alternate access providers* (AAPs), have brought fiber to customers that are "large enough" but not to smaller customers, residences, and educational institutions. By contrast, the cable TV companies have not yet brought fiber connectivity directly to their end customers. However, they have embarked on efforts to use fiber as far as the "node" in the neighborhood, from which traditional coaxial runs emanate. Those systems are known as HFC. Improved HFC networks support more bandwidth (currently up to 1 GHz; greater amounts in the future), use compressed digital video in both the server and the transmission plant, and even employ ATM.

One of the more sophisticated such networks to emerge is Time Warner Cable's Full Service Network. This network, under trial at press time, supports traditional cable TV distribution services, video on demand with instant access, interactive TV services, interactive games, access to long distance telephony, video telephony services, and *personal communication services* (PCS). A representative, reasonably sophisticated HFC cable TV network of today would support 450 MHz of bandwidth; it would have fiber spans no longer than 10 miles, although that can be extended farther with hub sites and regeneration; it would have approximately 5 total miles of coaxial cable per node, with a maximum of three amplifiers in cascade; each node would serve about 500 homes; and it would typically use the band 3–33 MHz for the upstream channel to support some level of interactivity. Table 1.3 provides information on alternative HFC systems. Compared with the all-digital and hybrid analog/digital HFC systems, the all-analog HFC system provides the lowest per-subscriber cost, especially for low "take-rates" (25–40%). Such a system, however, is the least sophisticated system in terms of supported services and, hence, in terms of revenue potential for carriers (typically these systems support only broadcast video). Also, the cost per subscriber increases considerably when the providers want to deliver a number of channels that necessitates two coaxial cables into each home. In the future it is likely that digitally modulated fiber (SONET-based) will be employed more extensively, particularly for *fiber-to-the-home* (FTTH) and *fiber-to-the-curb* (FTTC) architectures, especially to support ATM.

Table 1.3
Some HFC Architectural Alternatives

All-analog fiber/coax	Fiber from the headend or CO to a node in the consumer's proximity handling optoelectric conversion. Beyond the node, a coaxial bus distribution network is employed. Such a system typically provides a combination of analog broadcast video channels, in addition to a number of premium pay channels, selectable on a pay-per-view or subscription basis. This access method does not support the full complement of services of interest (e.g., video on demand). However, it can be used as a benchmark for economic comparisons, as it is very similar to the cable TV networks currently deployed.

Table 1.3 (continued)

Hybrid analog and digital fiber/coax	Fiber from the headend or CO, followed by a coaxial tree-and-branch distribution network beyond the node handling optoelectric conversion. Such a system typically provides a selection of both analog channels (basic and extended programming) and digital pay channels (premium, pay per view, and video on demand). At the home, customers access pay channels through digital setup boxes that demultiplex and decode the video signals and convert them to the analog format needed by the entrenched television set.
All-digital fiber/coax	This system entails fiber from the headend or CO, followed by a coaxial tree-and-branch distribution network beyond the node handling optoelectric conversion. The system supports a combination of broadcast and switched video channels in digital form to the subscribers' homes. At the home, subscribers utilize digital setup boxes to access all the video channels.

Dialup

Dialup access is based on traditional analog telephony methods. This so-called *plain old telephone service* (POTS) approach involves the use of modems to connect the user PC (owned by the user or provided by the corporation) to the remote server or host; it utilizes the analog public telephone network. Circuit switching implies that the communications channel is not dedicated 24 hours per day but must be brought on line when needed (via a process called *call setup*) and then taken down when no longer needed. Traditional modems have operated at speeds up to 19.2 Kbps and are now up to 28,800 Kbps; however, until recently, 9.6 Kbps has been more common (Table 1.4). That implies that the throughput across this type of link is fairly small; consequently, only a small number of users or short inquiry/response-like transactions can be supported. Since the link between the two points is not available on a dedicated basis, the user needs to dial up the information provider, as needed. File transfer is limited to relatively small files.

Video Dial Tone

VDT is an emerging information exchange and exchange access service that is being offered by LECs to support *video information users* (VIUs), including distance learning receivers and providers, in accessing video applications, as well

Table 1.4
Modem-Supported Speeds

Standard	Highest Data Rate (bps) (Without Compression)	Introduced	Increase in Data Rate Over Previous Standard
Bell 103	300	1960s	—
Bell 212 (v.22)	1,200	1978	4.0
V.22*bis*	2,400	1981	2.0
V.32	9,600	1986	4.0
V.32*bis*	14,400	1991	1.5
V.34	28,800	1994	2.0

as multimedia applications, offered by *video information providers* (VIPs). Each VIU household subscribing to a VDT service may be able to establish either a single session or multiple sessions, depending on implementation.[6] Each VDT session includes three types of information flows [9]:

- *One-way video distribution.* Initially, VDT services offered by the LECs will support one-way video. Video distribution may be point to point or point to multipoint, depending on the application.
- *Associated two-way data flows between VIPs and VIUs, if required.* Data communication requirements for VIU/VIP interactions differ, depending on the VDT application.
- *Signaling.* VDT applications require signaling because of the need to control the access and distribution of information. The user needs to select the appropriate VIP and the appropriate program from that VIP; signaling is integral to those functions.

Because of the required bandwidth, the implementation of VDT networks can be based on ATM over SONET. ATM can provide the switching function, while SONET provides the transmission functions. Through the ATM technology, only the channels of specific interest are switched into the user's access

---

6. VDT should not be confused with the cable TV industry's drive to enhance their systems, for example, via HFC. VDT is an FCC-sanctioned RBOC-provided system to allow access to multiple video service providers. As a delivery mechanism, VDT could use technologies such as HFC, fiber, enhanced twisted-pair, and so on

facilities. The backbone portion (from the serving CO to the VIP) could employ an OC-24 (1.2 Gbps) transmission system or an OC-48 (2.4 Gbps) transmission system. The access portion would need from 45 Mbps to 155 Mbps, to support, say, five 6-Mbps Motion Picture Expert Group-2 (MPEG-2) channels or five 21.5-Mbps *high-definition television* (HDTV) channels.

Six access technologies can support VDT services and are either considered or implemented by at least one RBOC. These solutions are [9]: ADSL, HFC, *fiber to the node* (FTTN), FTTC, *fiber to the building* (FTTB), and FTTH or fiber to the site.

Cable Television Solutions

Cable TV companies provide local communities with an increasing number of services. These services include access to educational TV programs, data over cable services, interactive TV services, and telephony. In addition to improved video delivery using HFC technology, data communication over the same cable is also being advanced. The key components of data on cable systems include:

- A cable modem, with PC adapter card. This modem uses subscriber data to modulate a 6-MHz carrier upstream to obtain bandwidth in the range of 1–10 Mbps. Currently, this type of modem tends to use proprietary encoding methods. It also receives and demodulates a 6 MHz carrier downstream.
- A frequency translator in the headend that takes upstream transmissions from customers and converts the transmission signal to a downstream broadcast channel. Multiple translators can support multiple independent LANs.
- A bridge or router in the headend that provides access to WANs.

ISDN

This information-access approach involves the use of switched *digital* facilities between the user and the information or education provider. ISDN provides end-to-end digital connectivity with access to voice and data (both circuit switched and packet) services over the same digital transmission and switching facilities. It provides a range of services using a limited set of connection types. ISDN supports a small number of physical *user-network interfaces*. The more well known of these interfaces is 2B+D interface, which provides two switched 64-Kbps channels, plus a 16-Kbps packet/signaling channel (144 Kbps total).

Other less-known ISDN interfaces include 1B+D, which are currently trialed by some LECs, and NB+D, where N is greater than 2 but less than 23.

Although many users now employ faster modems, operating at 14.4 Kbps, 19.2 Kbps, 38.4 Kbps with compression, and even 57.6 Kbps with even more compression, ISDN offers a more "comfortable" 64 Kbps, 128 Kbps, or even 384 Kbps throughput, affording adequate flexibility to the user. Some see a combination of ISDN and FRS as a viable way to bring in remote users and connect them to the information provider's network. ISDN can support data-oriented as well as video/audio-oriented distance learning applications.

Satellite Communication

Satellite transmission is used extensively in video broadcast in the United States and in many other areas of the world. Less developed countries and rural environments with dispersed populations or rugged terrain have found satellite systems to be the practical way to support their telecommunications needs. Geosynchronous satellites, which maintain a fixed position relative to the ground, have been used for commercial applications for over 25 years. They have seen increased deployment, starting in the early 1980s, as relatively small Earth stations became available. Typical applications include video broadcast, voice telephony, and wideband data transmission implemented on a point-to-point or a point-to-multipoint basis (using one-way transmission). The major potential for satellites is now for private network services on a point-to-point or full-mesh basis.

A satellite is a communication device that receives signals from a ground station, amplifies them, and broadcasts them to all Earth stations capable of viewing the satellite and receiving its transmissions. The satellite is an active radio relay, much like the radio relays used in terrestrial microwave communications: A satellite transmission begins at the Earth station, traverses the satellite, and ends at one or more Earth stations.

A satellite communication link involves three elements: the space segment, the transmission path element, and the ground segment. The space segment consists of the satellite itself, along with the supporting telemetry to keep the satellite in orbit. The transmission path is the radio spectrum used for communicating to and from the satellite; issues such as frequency, signal interference, modulation schemes, and protocols used to ensure proper transmission and reception (multiplexing and multiple access schemes that allow fair and efficient access to satellite channels) affect this portion of the overall system. The ground segment includes the Earth stations; issues such as placement, construction, and types of antennas used for different applications are part of the design of a satellite-based communication system.

Many distance learning systems currently employ satellites to achieve one-to-many transmission of the video program; the return audio path is generally supported over a dialup telephone link from the remote sites.

Very Small Aperture Terminals

Very small aperture terminals (VSATs) are a satellite transmission technology that was introduced in early 1980s. VSATs integrate transmission and switching functions to support on-demand links for point-to-point, point-to-multipoint applications. It is (generally) an asymmetric technology with a shared inband channel and (generally) a common outbound channel. VSAT Earth stations include the following elements:

- An antenna of small diameter (e.g., 1m);
- A radio-frequency power unit that provides 1–5W of power to support communications at bandwidth up to 1.5 Mbps over the transmission band (typically the Ku band at 12/14 GHz);
- A *master Earth station* (MES), linked via satellite to VSAT Earth stations, which performs routing and management functions. The MES can be either operated by a large company or shared by small companies subscribing to a facility operated by a service provider.

VSATs are used for data transmission and possibly for some compressed-video programming. Inband, the bandwidth is usually shared using random access techniques over a 64-Kbps or 128-Kbps channel; outbound, a *time division multiplexing* (TDM) approach is employed over a 768-Kbps or 1.544-Mbps channel.

1.9 BUYERS OF DISTANCE LEARNING SOLUTIONS

The applications and networking solutions introduced in this chapter have emerged to satisfy the diverse needs of the buyers of distance learning solutions. These buyers exist in urban, suburban, and rural areas and include distance learners such as corporate employees, high school students, and university students. They also include education providers such as universities, corporate training organizations, and public television education providers.

The appropriate distance learning application solution depends on the application linking the distance learners and the distance teachers. Part I (Chapters 1–5) of this book describes the requirements of the various constituencies and analyzes the applications driving the needs of the buyers of distance learning solutions:

- Chapter 2 analyzes K–12 learning applications.
- Chapter 3 analyzes corporate distance applications.
- Chapter 4 analyzes higher education distance learning applications.

In each of those chapters, the following approach is used:

- The problems facing the distance learning are identified.
- The distance learning applications by those problems are described.
- The needs of the end users and the telecom/datacom managers (or, more generally, program administrators) associated with each application are analyzed.
- The application and networking solutions that can potentially address those needs are described and evaluated.
- The strengths and weaknesses of the solutions are described.

Chapter 5 discusses many of the statewide initiatives underway at press time that have been undertaken to meet some of those needs. More intitiatives are expected in the future.

1.10 THE NATIONAL INFORMATION INFRASTRUCTURE AND DISTANCE LEARNING

At this time, the networking solutions needed by education providers and receivers are not uniformly available, nor are they uniformly affordable. The affordability and ubiquity of these services cannot be improved without the creation of a *national information infrastructure* (NII). This is a topic of intense debate and interest to the federal government as well as to state governments, universities, and network services providers. A number of initiatives are currently being carried out by these players. Part II of this book explores some early initiatives and the state of NII development.

1.10.1 The Role of State and Federal Governments

The federal government and the state governments play vital indirect roles in speeding the process of creating an NII as well as in education reform. These roles include:

- *Financial role.* The financial role of the federal and state governments refers to their efforts to finance and build statewide networks (as described in Chapter 5).

- *Regulatory role.* The regulatory role of the state governments refers to the efforts of the public utility commissions of the individual states as well as to those of the FCC to remove barriers to competition and to provide incentives to the expansion of the distance learning programs. The telecommunications reform legislation of 1996 opens the telecommunications marketplace to greater competition, which, many expect, will increase the quality of and reduce the cost of service.
- *User role.* The state and federal governments are large consumers of information technology.
- *Educator role.* The state and federal governments play an important role as advocates of distance learning. They also play a direct role in resolving education problems.

To play these roles effectively, the federal and state governments need to introduce regulatory and legislative reforms (workfare reform, education reform, and immigration reform). They can also provide financial incentives to education providers and receivers. In addition, they can get directly involved in the distance learning industry by introducing new distance learning technologies. These technologies can address many problems such as the education of welfare recipients and the structurally unemployed. These strategic moves on the part of the government are explored in detail in Chapter 6.

1.10.2   The Role of the Telephone Companies

Carriers and telephone companies play an active role in building the NII and in delivering a number of services that ride on the NII. The telephone companies view the NII as a tool to face these challenges and to differentiate themselves from their existing and emerging competitors. To create the NII, the LECs are taking the following steps:

- They are upgrading their network infrastructure to support broadband applications.
- They are building broadband testbeds to stimulate the demand for broadband services (e.g., Pacific Bell's California Research and Education Network, or CALREN), including distance learning.
- They are bidding for—and sometimes winning—contracts to support state government efforts to build statewide networks.

Chapter 7 explores the current and potential NII/distance learning initiatives of the telephone companies.

1.10.3 The Role of the Cable TV Companies and the Alternate Access Providers

Two closely related groups, the cable TV companies and the AAPs, are the main competitors of the traditional telephone companies in the creation of the NII and in the delivery of distance learning services. Their roles include building joint regional cable TV–AAP networks to provide an increasing number of services, including data over coaxial cable and access to educational TV programming. AAPs are now receiving equal status to the LECs in a number of states (e.g., Teleport Communications Group in Connecticut). While many AAPs are allied with the cable TV companies, a few companies are not. These independent AAPs are building MANs and offering services, primarily to business users, that include private-line services, LAN interconnection services, and, increasingly, switched data services.

Most AAPs are evolving along two parallel tracks: the traditional AAP—a fiber network providing private lines and special access in a downtown central business district—is becoming an endangered species in the view of some observers [21]. AAPs are becoming either national AAPs or what have been called competitive LECs (CompLECs, also known as CLECs). National AAPs will continue to do what AAPs have done over the years: serve telecommunications-dependent businesses, which need the operational and strategic security that is accrued from having a diverse source of supply for critical local telecommunications services. But it is no longer enough for an AAP to be a diverse source only for so-called "dumb" services such as private lines and special access. Hence, the more progressive AAPs (e.g., Teleport Communications Group) are providing "intelligent" enhanced local switched services such as ISDN, Centrex, and switched access as well as new data services, including Internet access, ATM, and FRS. Chapter 8 provides an analysis of the roles of the AAPs and cable TV companies and an assessment of their contributions to the creation of the NII; that chapter also examines the critical success factors shaping the ability of those groups to contribute to the emergence of the NII.

1.10.4 The Role of the Internet Community

The Internet will continue to play a vital role as a key component of the U.S. (and the global) information infrastructure and as a major player in the distance learning industry. Its continued growing role is attributed to: (a) its broadband infrastructure, which is migrating from a DS3 backbone to an ATM/SONET backbone; (b) its growing role in supporting the video/videoconferencing needs of K–12 students, in addition to its traditional role in supporting the collaborative learning and research efforts of universities and corporations; and (c) the ever increasing number of participants in the Internet. Chapter 9 provides an analysis of the role of the Internet in supporting the NII and distance learning and the strengths and limitations of its current efforts.

1.11 CASE STUDIES

The demand for IDL solutions will be shaped by the ability of the IDL providers (including application solution providers and network solution providers) to meet the needs of buyers. An analysis of the success of network service providers in satisfying the needs of buyers cannot be accomplished without conducting case studies. Case studies describe the experiences of buyers in implementing new IDL technologies, the problems they encountered in implementing those technologies, and the potential improvements they would like to see in IDL services. This is the topic of Part III (Chapters 10–12), which provides three specific case studies drawn from the experience of IDL practitioners and providers. The case studies explored in Part III cut across a number of the IDL groups described in Part I, including universities, colleges, high schools, corporations, and service providers.

References

[1] Hudspeth, D. R., and R. G. Brey, *Instructional Telecommunications, Principles and Applications*, New York: Praeger,1986.

[2] Gayeski, D. M., *Multimedia for Learning Development, Applications, Evaluation*, Englewood Cliffs, NJ: Educational Technology Publications, 1993.

[3] B. Duning, B., Van Kekerix, and Zaborowski, L., *Reaching Learners Through Telecommunications*, San Francisco: Jossey-Bass, 1993.

[4] Harasim, L. (ed.), *Online Education*, New York: Praeger, 1990.

[5] Hiltz, R. S., *The Virtual Classroom*, Norwood, NJ: Ablex, 1995.

[6] Moore, M., ed., *Contemporary Issues in American Distance Learning*, Oxford, Eng.: Pergamon, 1990.

[7] Watkins, B., and S. Wright, *The Foundations of American Distance Education*, Dubuque, IA: Kendall/Hunt, 1991.

[8] Zigerell, J., *The Users of Television in American Higher Education*, New York: Praeger, 1991.

[9] Minoli, D., *Video Dialtone Technology, Approaches, and Services: Digital Video Over ADSL, HFC, FTTC, and ATM*, New York: McGraw-Hill, 1995.

[10] Hezel Associates, *Educational Telecommunications, State By State Analysis*, Syracuse, NY, 1994.

[11] Sony Distance Learning promotional literature, 1994.

[12] Minoli, D., "Distance Learning Applications, Broadband Networking," *DataPro Report*, 1015BBN, November 1993.

[13] Find/SVP, *The American Internet User Survey*, New York, 1995.

[14] Minoli, D., *Enterprise Networking: Fractional T1 to SONET, Frame Relay to BISDN*, Norwood, MA: Artech House, 1993.

[15] Minoli, D., *Telecommunication Technologies Handbook*, Norwood, MA: Artech House, 1991.

[16] Minoli, D., and M. Vitella, *Cell Relay Service and ATM for Corporate Environments*, New York: McGraw-Hill, 1994.

[17] Minoli, D., and G. Dobrowski, *Signaling Principles for Frame Relay and Cell Relay Services*, Norwood, MA: Artech House, 1994.

[18] Minoli, D., and R. Keinath, *Distributed Multimedia Through Broadband Communication Services*, Norwood, MA: Artech House, 1994.

[19] Minoli, D., *First, Second, and Next Generation LANs*, New York: McGraw-Hill, 1994.

[20] Eldib, O., and D. Minoli, *Telecommuting*, Norwood, MA: Artech House, 1995.

[21] TCG press release, CompLECS & Universal Service Assurance, Staten Island, NY, August 1994.

The Distance Learning Needs of K–12 Schools

2

K–12 schools are facing major problems that, according to many observers, threaten to erode the foundation of the U.S. education system. The goal of this chapter is to explore some of these problems and to describe the IDL applications and solutions that can be leveraged to solve them. To accomplish that goal, the chapter first identifies the educational challenges facing the K–12 schools and then proposes a communication model that identifies the existing and potential IDL communications links that K–12 schools need to establish. After that, the chapter analyzes the applications driving the needs for the establishment of those links. The chapter further describes the applications and networking solutions on which K–12 schools can rely to meet their distance learning needs and the strengths and limitations of each solution. The chapter concludes by comparing solutions.

This chapter is targeted at K–12 school administrators, students, and teachers who need a better understanding of the IDL solutions that can address their needs. It is also targeted at the network service providers that are searching for specific IDL business opportunities within the K–12 market. (Chapter 9 provides a case study for implementing high-performance IDL systems in K–12.)

2.1 THE CHALLENGES FACING K–12 STUDENTS

K–12 schools in the United States are faced with the challenge of providing equitable, far-reaching, practical, and high-quality educational services. Meeting that challenge is important so K–12 schools can produce graduates who can compete effectively with graduates from schools in other industrialized countries. This challenge cannot be met unless K–12 schools face the following problems head on:

- *Illiteracy.* U.S. schools graduate 700,000 functionally illiterate students every year. These graduates lack the basic skills that enable them to function effectively as part of the U.S. workforce. Basic skills expected in industry include reading, writing, mathematics, and interpersonal communications.
- *High dropout rate.* About 700,000 students drop out before receiving a high school certificate [1]. Some of these students tend to become a burden on the American society, costing the government funds associated with crime prevention, welfare, and healthcare.
- *Relatively low skills.* U.S. students who do stay in school often cannot compete effectively with students from other industrialized countries. Published data about international tests indicate that children in the United States rank at the bottom of most international tests, behind children in Europe and Southeast Asia. Students in other countries are more skilled in computational activities, which are of increasing importance for an effective and flexible labor force.
- *Weak link between education and real life.* The U.S. education system is not closely linked to the needs of the U.S. labor force. Unlike Germany, which has a strong and effective apprenticeship program, the United States lacks a program of the same quality. That creates two problems: (a) the U.S. labor force cannot compete effectively with the other industrialized nations, and (b) many students are not motivated to stay in school because they do not see a direct relationship between school and real life.
- *Growing shortage of science and math teachers.* According to some industry forecasts, by the year 2000, the number of public school students could reach 44 million, an increase of nearly 10%. At the same time, only one million teachers will be available to fill two million positions.
- *Physical isolation of K–12 schools.* Another major problem facing the U.S. education system is that traditional K–12 classrooms are, to a great extent, physically isolated from other academic environments, such as universities, libraries, and museums [2]. The traditional classroom is also isolated from practical sources of knowledge, such as corporations and R&D institutions. Most classrooms lack the electronic means (e.g., modems and telephone lines) to access those environments.
- *Inequity of information distribution.* Another challenge facing the U.S. education system is education inequity. Rural communities are becoming increasingly isolated as the number of students and the resources available to them decline. One lacking resource is teachers. As a result, those communities are becoming "information poor" relative to more affluent suburban communities. The inner cities are also facing the same problems. Their information poverty is a direct extension of their financial poverty.

- *Increasing budgetary pressures on K–12 school administration.* K–12 schools, particularly public schools, are under increasing financial pressures, which can be attributed to the growing financial deficits of local and state governments. Budget deficits are forcing government agencies to reduce funds allocated to K–12 schools.
- *Lack of educational programs to support gifted students.* Gifted students have needs that extend beyond those of average students. They often want to acquire more knowledge, such as additional languages and college courses, so their full potential can be utilized.
- *Lack of adequate teacher training.* To keep up with the latest advances in their areas of expertise, teachers need more systematic and extensive training.

The news, however, is not totally bleak. First, students who are motivated to achieve do very well in an economy that has plenty of room for innovation, entrepreneurship, ventures, startups, and hard workers. Second, the United States continues to lead in the areas of computer design, software development, postsecondary education, research, new applications of technology, aircraft design and manufacturing, and genetic sciences. Third, in my opinion, industry in general is shedding crocodile tears in terms of the supposed lack of available talents. In fact, there are more mathematicians, chemists, electrical engineers, thinkers, and academicians that industry can absorb; many workers are underemployed in terms of both their chosen profession and their formal training. Some workers feel that, although they may be working at premiere companies, perhaps only a fraction of their analytical capacities are put to use by their employers. For example, a mathematician who received a college degree in 1985 could still be solving mathematical problems in the year 2040.[1] In my opinion, what corporations lament is their inability to find people of the highest possible caliber at entry-level salaries for an indefinite period of time. Effectively they may be implying that after employing people right out of college for 5 years, they would want to encourage those employees to move on, so the corporations can recruit "fresh blood" again at entry-level salaries. Hence, the so-called lack of appropriate talent may be greatly exaggerated by corporations: the talent is there, but youth is passing.

2.2 COMMUNICATIONS LINKS OF K–12 SCHOOLS

The problems facing K–12 schools can be attributed to two key drivers: (a) *inadequate resources* (proficient teachers, adequate funding levels, adequate courses) and (b) *geographic isolation* (isolation of rural communities and inner

---

1. This assumes that someone born in 1960 could be productive into his or her 80s.

cities from information). Interactive distance learning solutions can address many of those problems by establishing electronic links between the K–12 resources (the teachers and the students) and the various sources of learning. However, K–12 school administrators cannot hope to maximize the benefits of instituting a distance learning program without developing a conceptual framework of the information links that the key players in the K–12 school community need to establish.

Figure 2.1 provides a framework of the electronic links that K–12 schools need to establish with various sources of learning. Sources of learning include other K–12 schools, universities, libraries, museums, public TV broadcasters, and students' homes. These links support the following IDL applications, among others:

- Remote K–12 education;
- Collaborative learning, such as remote college education;
- Access to museums;
- Access to libraries;
- Access to public broadcasting service providers;
- Access to students' homes;
- Access to the Internet.

Each of these applications is described next.

2.2.1 Remote K–12 Education

K–12 schools need to establish remote education links with other K–12 schools that may be located in rural communities or in the inner cities. Remote education links can reduce the isolation of rural K–12 school students by providing them with access to special teachers. Remote education links are also necessary to provide disadvantaged students in inner cities access to teachers that are not available in the students' districts.

For remote education links to be effective, they need to be based on a high-quality video/videoconferencing solution. Students have grown accustomed to visual communications in the home, typified by *National Television Standards Committee* (NTSC) quality of motion picture resolution. Consequently, quality video and audio are essential to maximize the benefits of the remote learning process.

2.2.2 Collaborative Learning

Collaboration through advanced technologies, such as videoconferencing, is an emerging trend in K–12 schools. Experts in the field of education believe that through advanced technologies, such as data conferencing and WANs, students

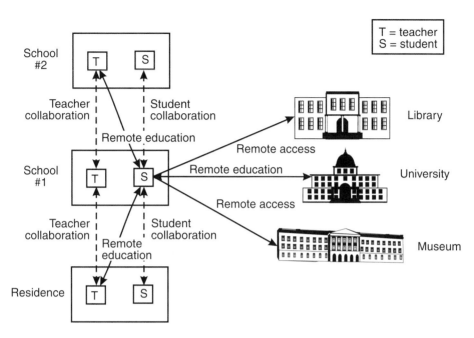

Figure 2.1 Distance learning links of K–12 schools.

learn more effectively. This kind of learning entails the process of being engaged in active exploration, interpretation, and construction of ideas with other students who share the same school or the same district. Such collaboration enables educators to share and analyze scientific data; have extended discussions with their peers; and collaborate in producing papers and journals. At the K–12 level, collaboration lets students share projects, exchange ideas, share homework, and publish newspapers or newsletters.

For collaborative learning to be effective, educational experts believe that schools have to change in the following ways:

- Collaboration becomes more effective as the number of teachers and students participating in a specific collaborative event decreases [3]. The fewer the number of students that participate in a given collaborative effort, the closer their ties with their teammates, which encourages them to be more confident and to become more engaged in the learning process. Some educational experts believe that schools should have 500 or fewer students and that teachers should have no more than 40 to 50 students whose work they know well.
- Effective collaboration among students and teachers belonging to the same team requires more effective scheduling for group work, for reflection and

revision, and for thoughtful assessment of learning. Effective collaboration also means that interactions among members need to be expanded beyond school hours.

- Technologies have to be in place to support collaborative learning. These technologies include groupware, video and videoconferencing, institutional LANs, and community MANs or WANs. The extent of these technologies needs to be extended beyond the labs and into the classrooms.

- K–12 school administrators need to change the traditional mindset, which maintains that students learn best in isolated, intensive, "bite-sized" encounters with subject matter and with the teacher. That mindset needs to be replaced with the more modern view that students learn best when they work in a collaborative mode.

- High schools need to establish quality videoconferencing links with colleges and universities. This experience would be particularly valuable to gifted students who want to prepare themselves for college courses. K–12 school links with universities would also be valuable to teachers who want to upgrade their skills either by taking courses or by completing a degree in their field of expertise.

2.2.3 Access to Museums

Students, particularly those in suburban and rural areas, would value the experience of taking electronic field trips to museums through video/audio conferencing links. These electronic field trips would be helpful to students with an interest in a variety of fields, such as science, technology, and history. Electronic field trips could be either a substitute for or a complement to physical field trips. In a project undertaken by the author, the American Museum of Natural History recently deployed a broadband-based campus network to support multimedia, World Wide Web (WWW) pages, distance learning for lectures, and dissemination of collection information (specifically anthropology), a portion of which has been digitized.

2.2.4 Access to Libraries

Access to libraries is an important data-oriented IDL application. By retrieving image-based information from digital libraries, students can save the time associated with physically visiting a library building. Digital libraries can also provide students with keyword access to a wider range of information resources that might not be available in their local libraries. A system such as this needs communication links as well as supporting software tools. For a data solution to be effective in supporting students' need for access to a library, it has to be easy to use. Preferably, that means the user interface should be graphical as opposed to an antiquated UNIX-like interface.

2.2.5 Schools Links With Homes

Schools need to establish stronger electronic links with students' homes to support parents/teacher interactions after hours: Studies show that parents would like to be more involved in the education of their children. The busy schedules of parents, particularly working parents, limit the possibility of face-to-face interactions with their children's teachers. In addition, many parents are becoming increasingly computer literate; consequently, they would welcome the opportunity to establish data communications links with their children's teachers. That would enable them to exchange views regarding areas of improvement in the students' performance.

2.2.6 Access to Internet Information

Many institutions publish information on servers connected to the Internet, including library information, research information, Congressional bills, NASA pictures, weather maps, and many other kinds of information. Certainly, this information can be put to effective use by educators and students alike.

2.3 THE NEEDS OF K–12 IS ADMINISTRATORS

In implementing IDL application solutions, program and telecom/datacom managers should take into account the following factors:

- *Low cost.* Telecom/datacom managers of school districts, particularly public schools, are sensitive to cost, considering the continuing trends for tight budgets of their school districts.
- *Geographic coverage.* Telecom/datacom managers of school districts seek application solutions that support the geographic range of the applications they support. As discussed earlier, K–12 schools may be located in urban areas or suburban areas. They may also be located in rural areas, where fewer telecommunications services are available. The wide-area reach includes metropolitan community-level facilities (i.e., libraries, court houses, local industrial laboratories, and medical facilities), access to Internet points of entry, and other wide-area access.
- *Support for installed base.* Telecom/datacom managers value solutions that can build on their limited installed base of information systems. Schools generally are deficient in terms of the installed base of computing and telecommunications technology. Few schools have installed telephones to all classrooms in a building. Few schools have as many computers for students as they need. Many of the PCs that K–12 schools own are

donations from major computer manufacturers, such as IBM and Apple Computer, and may be of older vintage.

- *Ease of use and administration.* Telecom/datacom managers value application solutions that are easy to administer and to use. This is a particularly important criterion to K–12 schools, considering the limited exposure of the K–12 teachers to telecom/datacom technologies.
- *Fixed versus variable costs.* K–12 schools prefer fixed communication and information access charges so they can better manage their budgets.
- *Intraenterprise versus interenterprise communications.* The appropriateness of the application solution may depend on the extent of the needs of telecom/datacom managers for intercompany versus intracompany communications.
- *Access methods and speeds.* Telecom/datacom managers need different access methods and speeds in support of application solutions. Those requirements are shaped by the installed base as well as by cost factors. The range of access methods and speeds required by telecom/datacom managers in support of application solutions may differ from one school to another. Access speeds may range from 1.2 Kbps to 1.5 Mbps. Quality video requires even more (i.e., 6 MHz for analog video, 6 Mbps for digital *standard-definition television* [SDTV] video, and 21.5 Mbps for HDTV).
- *Outsourcing.* Compared to universities and corporations, K–12 schools have limited telecom/datacom human resources. Consequently, they value application solution providers that enable them to outsource as many telecom/datacom and communications functions as possible [4].

The financing of IDL always represent a point of concern for planners. The USDLA publishes a funding guide. IDL funding sources include foundations, technology companies, federal grants, state grants, school districts, and colleges. Foundations are now setting aside more funding for the application of technology to education. In recent years, however, technology and telecommunication companies have become more sparing in their "demonstration" funds.

2.4 DISTANCE LEARNING APPLICATION SOLUTIONS AVAILABLE TO K–12 SCHOOLS

A number of application solutions are available to meet the needs of K–12 schools for IDL applications, including:

- The Internet;
- Value-added network providers;
- Groupware;
- Educational TV programming (one-way video solutions);

- Two-way videoconferencing, including continuous presence.

Each of these solutions is described next.

2.4.1 The Internet

As a data communications solution, the Internet can support the following K–12 IDL applications.

- *Access to libraries.* The Internet enables K–12 students and teachers to access digital libraries located throughout the United States and the world. They can also access online databases such as Dialog, Dow Jones, and Lexis/Nexis. The Internet also provides access to an online software library, with millions of public-domain shareware. Access to information resources on the Internet has been made much more user friendly through growing services, such as the WWW. This new service makes information resources on the Internet accessible to "regular" users who are not experienced in UNIX-based commands. Consequently, the WWW makes the Internet a more useful tool to new groups of information users, such as K–12 students, teachers, and students' parents. These groups can access WWW from school buildings or their homes.
- *Collaborative learning.* Students can collaborate with each other through services, such as *file transfer protocol* (FTP), e-mail, and whiteboards. Through FTP services, students can share files with each other. Through e-mail, students can exchange notes about homework. Through whiteboards, students and teachers can share their notes in real time with viewers over a wide area. This tool makes the dialog among educators and students more interactive and more effective, enhancing the collaborative research and learning process.
- *School links with students' homes.* Through the services mentioned in this list, a student's parents can exchange messages with the student's teachers from their homes or even from their place of works.

The Internet is also emerging as an entry-level videoconferencing solution in support of the following IDL applications: (a) remote education links among K–12 schools, (b) remote college education, and (c) electronic field trips to museums. The videoconferencing capabilities of the Internet are being made possible through *National Science Foundation* (NSF)–initiated activities, such as the Global School House program. That program demonstrated that the Internet can be used to allow students all over the world to establish low-end interactive videoconferencing links with each other and with teachers, scientists, and national and international leaders. The first students to participate in the Global School House project were fifth- to eighth-graders from Oceanside, California;

Knoxville, Tennessee; Arlington, Virginia; and Hampton, England. Students from those schools worked together on a study of the environment. Over the course of a six-week period, students conducted independent research on groundwater pollution and its source in their communities. In communicating with each other, students relied on Apple Macintosh computers running CU-SeeME software developed by Cornell University. That software enabled the students to see each other on a video monitor. Students also relied on FrEdmail software, which enabled them to communicate with each other regarding the progress of their work.

The application of the Internet as a videoconferencing solution will be of increasing value to the K–12 schools for the following reasons:

- It will provide students with international exposure, enabling them to interact with other students throughout the world.
- It will enable students to enhance their learning skills by getting immersed into a topic (such as the environment) and by jointly exploring the issues associated with that topic.
- It provides telecom/datacom administrators with a potentially low-cost solution.
- It prepares students for participation in the global economy.

The Internet solution is beneficial not only to K–12 schools but also to telecom/datacom administrators of school districts for the following reasons:

- *Support by network service providers.* Access to the Internet is increasingly supported not only by the traditional Internet providers but also the AAPs, the LECs, and the IXCs.
- *A high-performance backbone.* The NSF backbone is migrating from a DS3 (45 Mbps) platform to a higher performance network, based on ATM and SONET (155 Mbps–10 Gbps). (For more information about the Internet infrastructure, refer to Chapter 8.)
- *Affordability.* The Internet remains one of the least costly approaches to providing interconnection, delivering on the NII promise of an affordable networking solution [5]. The Internet cost to the user may range from $1 to $10 per hour in addition to telephone charges. The low cost of access to the Internet enables large companies as well as small companies to establish business communications links with potential buyers and suppliers.
- *Support for interdistrict and intradistrict networking solutions.* The Internet is appropriate for establishing intradistrict IDL communications links as well as links among districts and between a district and a university.
- *Geographic coverage.* The Internet has a global geographic coverage.

- *Ease of use and administration.* The Internet is increasingly easy to administer and to use. This is particularly helpful to telecom/datacom managers in supporting IDL need of K–12 teachers, who have limited exposure to telecom/datacom technologies.
- *Access methods and speeds.* Telecom/datacom managers can access the Internet through multiple access solutions, which are offered by the telephone companies and cable TV providers (access methods are discussed in more detail toward the end of this chapter).
- *Outsourcing.* By relying on integrated service providers, which offer application as well as networking and outsourcing services, the telecom/datacom manages can reduce their administrative cost [4].

All these factors make the Internet an appropriate IDL solution to K–12 school administrators. The Internet, however, cannot be considered a total solution to K–12 school administrators for the following reasons:

- *Lack of support for rural communities.* While the Internet is widely available, the Internet has not yet reached rural communities in terms of local access nodes (remote dial to a metropolitan access node, however, is possible).[2]
- *Lack of information filters.* The wide range of information resources available on the Internet is not only one of its strengths but also one of its weakness. This is because while most information available on the Internet enables students and educators to learn, explore, and collaborate, some information available on the Internet is literally destructive. A case in point is the K–12 student who was injured recently while attempting to make a bomb, having accessed information on the Internet. Also, some students become "addicted" to the activity known as "surfing the net."
- *The Internet as a public forum.* Like talk radio, the Internet has become a public forum where ideas are generated, discussed, and debated. Some of these ideas, however, spread negativism and cynicism. This may be particularly dangerous to members of the younger generation, who are trying to sort out the increasingly complex world in which they live.
- *No guaranteed performance.* The K–12 school administrators need to consider that they cannot be guaranteed a certain throughput across the Internet nor a consistent reliability level. The Internet, while serving thousands of organizations and millions of individuals, lacks any mechanism for reserving bandwidth. In addition, the Internet is made up of many networks, so the establishment of common reliability levels requires complex interactions among a large number of Internet providers. This is a particularly important issue to universities and corporations.

---

2. Many new ventures are now working on developing inexpensive local dial access technologies.

The efforts of schools to join the Internet community will be limited by the availability of the resources necessary to lease WAN lines to the point of access and to purchase computers and software.

2.4.2 Other Providers: Value-Added Network Providers

In addition to the Internet, commercial online services are available to educators, students looking for information for homework activities, and parents looking to help their children with homework. The major consumer online services are, in order of market share: Prodigy (a joint IBM-Sears venture), CompuServe (a division of H&R Block), America Online, GEnie (a division of General Electric Information Services), and Delphi. A gamut of other providers is available. These services counted about 5 million subscribers in 1994, and the number of subscribers is forecast to grow at a compound rate of 35% per year, to 35 million in the year 2000. In spite of that growth, only 9% of households with PCs access online services. Continued sales of PCs and modems will provide ample opportunities for information providers. In addition, Microsoft is including an online service with the 1995 release of Windows; the software will eventually support online banking, financial information, travel services, chat bulletin boards, and electronic magazines [14].

With the market opportunities afforded by the commercialization of the Internet, a considerable number of new entrants appear ready to approach the online services market. The largest players in traditional media are going online, from magazines, newspapers, cable programmers, and game developers to the film and recording industry. Content-based companies are going online for three basic reasons: (a) the relatively low cost of entry; (b) the assurance of finding a ready-made market from one of the major operators such as Prodigy and America Online; and (c) the evolution of the PC as a household multimedia device, which offers, in the view of proponents, more assurances in the near term than the various interactive-TV solutions that have been advanced thus far in the "convergence" industry [6].

Interactive multimedia services, such as access to video databases, video mail, and videoconferencing, are emerging at the technology level. However, "killer applications" in multimedia have yet to be discovered, and the online services industry is debating what approach will be required to deliver higher bandwidth and more costly consumer-oriented multimedia services. The online services that are thriving are those that have facilitated the formation of communities of interest and have enabled those users to communicate in an unimpeded manner. "Chat rooms" and bulletinboards are commercially successful applications found on narrowband networks.

Digital information services have been used in the business sector for a number of years, mostly for information retrieval and distribution. The trend now is for consumers to begin using such services: The availability of communication services at relatively low rates, particularly when measured in constant dollar terms, and the embedded base of home computers have sustained the growth experienced by services like Prodigy, CompuServe, GEnie, and America Online. In addition, the emergence of GUI software that incorporates "intelligent agents" is improving ease of use, stimulating increased penetration. The next evolution is perhaps in the IDL market.

2.4.3 Groupware Solutions

Groupware is a data-oriented IDL solution that provides K–12 schools with a partial substitute to the Internet solution. While the Internet is a public solution, the groupware can be implemented as either a private solution or a public solution. Groupware refers to software that can support at least one of the following IDL applications:

- Electronic messaging;
- Data conferencing;
- Messaging gateways.

As a data communications solution, the groupware solution can support the following K–12 IDL applications:

- *Access to libraries.* The groupware solution enables K–12 students and teachers to access public as well as private databases, by establishing gateways with the Internet.
- *Collaborative learning.* Students can collaborate with each other through groupware services, such as messaging and computer conferencing. Through messaging services, K–12 students can submit their homework to teachers via e-mail, leave messages, or inquire about specific problems they are encountering in completing a homework assignment. Groupware computer conferencing applications enable K–12 schools students to interact with other students and teachers in concurrent, collaborative ways. Through groupware capabilities, students can look at, modify, and approve spreadsheets and other documents by viewing windows that contain a "slate" or a "whiteboard" on their PCs and making individual contributions as though they were using different colored "pens." (Chapter 10 discusses the use of groupware, Lotus Notes in particular, in support of NYU's Virtual College.)

Groupware is also emerging as a videoconferencing solution, enabling it to support the following IDL applications: (a) remote education links among K–12 schools and (b) remote college education. Through a (small) window on their desktop workstations, K–12 students can establish a videoconferencing session with IDL providers [7].

The groupware solution is beneficial not only to K–12 schools but also to telecom/datacom administrators of school districts for the following reasons:

- *Low cost.* In implementing groupware, telecom/datacom managers incur a nonrecurring cost. This cost is either on a per-user or per-server basis.
- *Multiple applications.* Groupware supports multiple applications, including computer conferencing, electronic messaging, and messaging gateways. These are all applications on which interactive distance learners and receivers can rely in communicating with each other.
- *Ease of use.* Multifunctional groupware solutions are based on GUIs, which makes it easy for IDL receivers and providers to use. Consequently, groupware solutions are appropriate for all groups of IDL receivers and providers.
- *Training and consulting.* The leading providers of multifunctional groupware provide their buyers with training and consulting as part of their product packages.
- *Support for interdistrict and intradistrict solutions.* The groupware solution is appropriate for establishing intradistrict IDL communications links as well as links among districts and between a district and a university. The interdistrict links can be established through public network service providers, such as AT&T, which offers a Lotus Notes–based public groupware solution.

These factors make groupware an appropriate IDL solution for K–12 school administrators. Groupware, however, cannot be considered a total solution for the following reasons:

- As a public networking solution, groupware support currently is limited to a few major IXCs, such as AT&T and MCI.
- Groupware is predominantly implemented as a private application solution.
- The cost of selecting, implementing, and upgrading the groupware solution is not trivial.
- The implementation of groupware by K–12 schools requires the availability of technical resources to select, purchase, implement, operate, and maintain a groupware solution.
- The implementation of groupware by K–12 schools requires the availability of an information systems infrastructure, including PCs in each classroom.

These limitations of groupware solutions are likely to slow the growth of groupware in the K–12 market.

2.4.4 Educational Television Programming

K–12 schools can access a number of educational programs by subscribing to basic cable TV services. This has been, until now, a one-way video/audio solution that enables students (and teachers) to access educational programming about computers, communication, multimedia, software, and related matters. An example of an educational program is the Jones Computer Program, through which students can watch cultural lectures, debates, and symposia. Another example is the Horizons Cable Network, which was scheduled to begin in 1995. A third example is the Learning Channel, which provides general education and enrichment programming. A fourth example is CNN, which provides news coverage 24 hours a day. Yet another example is C-Span, which provides live coverage of Congressional hearings, debates, and speeches.

Educational TV programs provides K–12 schools with several benefits:

- Exposure to a wide range of news, commentaries, lectures, and history, which expands students' horizons and reduces the physical isolation of the classroom. It also enables students to keep in touch with real life.
- Students, who grew up with TV as a key source of fun, are conditioned to absorb educational programming delivered to them through TV.
- Educational TV programs are not costly to K–12 schools, which have access to basic TV services.
- Television educational programming provides students with information about a wide range of topics.

Television-based educational services do have the following limitations, however, as sources of IDL for K–12 schools:

- The services may not be available to K–12 schools, particularly those located in rural areas. (Some programs, however, can be accessed through satellite reception by a private Earth station.)
- These programs lack interactivity, although that is changing as many TV show hosts add question-and-answer segments to their shows (students call the hosts on an 800 line).
- PBS-based programs are increasingly faced with decreased funding from Congress as well as from state and local legislatures.

2.4.5   The One-Way Video/Two-Way Audio Solution

One-way video/two-way audio solutions are probably the most prevalent in supporting the following K–12 school IDL applications: (a) remote education links among K–12 schools, (b) remote college education, and (c) electronic field trips to museums. These solutions are made possible through the efforts of satellite providers and public education providers. For example, in New Jersey, 45 schools have signed on to a satellite system called the Satellite Education Resources Consortium, which is a nonprofit partnership between public broadcasters and 23 state departments of education. The cost to individual school districts amounts to $3,000 for a satellite antenna and about $500 per student per course [8]. Through the satellite system, students have been able to take courses, including Russian, advanced-placement economics, and world geography, from teachers who are hundreds of miles away. The satellite system also provides university training to teachers. For example, the Stevens Institute of Technology provides teachers with a mathematics curriculum.

The one-way video/two-way audio solution provides students with the following benefits:

- *Support for gifted students.* The satellite system makes it possible for a school to offer courses it could not otherwise provide because of limited student interest.
- *Interactivity through audio links.* Through the audio links, students participating in a satellite-based course can establish friendships with other students around the country and thus expand their horizons beyond the limited cultural experiences of their local environments.
- *Pleasant experience.* For students who have grown up with TV as a key source of entertainment, learning though a one-way video solution can be enjoyable.

The one-way video/two-way audio solution is beneficial not only to K–12 schools but also to telecom/datacom administrators of school districts for the following reasons:

- *Large number of satellite-based providers.*
- *High-quality picture.* Analog-based satellite systems deliver high-quality pictures. New digital methods such as compressed digital video (for both HDTV and SDTV) are now being deployed (see Chapter 11).
- *Affordability.* The cost of one-way transmission is much less than the cost of hiring a teacher.
- *Support for interdistrict and intradistrict networking solutions .*

- *Geographic coverage.* The satellite-based one-way video solution has a broad geographic coverage.
- *Support for rural communities.* Satellite systems can solve the problem of reaching rural communities.

All these factors make one-way video/two-way audio an appropriate IDL solution for K–12 school administrators. This solution does, however, have limitations. The major problem for students with one-way videoconferencing systems is the accessibility of teachers. A Japanese IDL teacher indicated that one of his major problems with being remotely located is that he "cannot tell the student to please come and see me after class." Another problem with this solution is the lack of interactivity. While students can see the teacher, the teacher cannot see the students. Consequently, the relationship that develops between the student and the teacher through face-to-face interactions cannot be duplicated through the one-way video solution. A third problem is scheduling: In many cases, the daily transmission hours are often at odds with a school's class schedules.

The author recently taught a five-site distance learning course using a corporate videoconferencing system. Although the system was two-way, the course was inefficient compared to a classroom rendition of the same class that the author gave about 20 times at NYU. He ended up covering only 50% of what he normally would have covered. First, the whiteboard could not be used because the remote sites could not see the writing. Second, the viewgraphs the author had employed over the years could not be seen on the remote 19-inch monitors, even though a document camera was used. Even after abandoning all text-based viewgraphs, the author could not use the ones with diagrams because the encoding algorithm gave rise to a blurring situation when he pointed to specific items on the chart. In addition, to the extent that the document camera was used, the author had to constantly play with buttons to switch between the document camera and a view of himself; he also had to play with buttons to cycle through the remote locations. In addition, he had a problem because the local audience could not have the benefit of overhead viewgraph projection and was limited to seeing on a local monitor whatever the remote locations saw. The author had to constantly put in questions such as "What does White Plains think of that? How about Syracuse?" to make the remote site feel part of the group. And because the encoding delay was about 2 seconds, the time before a question was answered was 4–5 seconds, which caused constant tripover. Such problems are even more serious in one-way video/two-way audio.

K–12 school administrators cannot consider the one-way video/two-way audio solution a total solution because it cannot support several IDL applications, including collaborative learning and access to libraries.

2.4.6   Continuous Presence

Continuous presence represents the high-end approach to IDL. With continuous presence, the instructor has control of multiple remote and local cameras that can be panned and zoomed. The delivered signal includes voice, video, and data. Each student may have a PC and directly receive the instruction, including individualized information, tests, assignments, and so on. Each classroom also has a projection screen with a quad-split picture of the instructor and up to three other classrooms, to give a sense of being "present."

The continuous-presence solution can support all the applications that the satellite-based one-way video/two-way audio solution supports, including remote college education, remote education links among K–12 schools, and electronic field trips to museums. In addition, this solution can support collaborative learning, complementing the groupware solution. The continuous-presence solution offers an additional benefit over the one-way video/two-way audio solution: interactivity. With continuous-presence solutions, all the participants can see each other, enhancing the collaborative learning process. The quality of the continuous presence picture, its affordability, and its geographic presence depend on the networking solution provided. Figure 2.2 depicts an ATM-based solution. Other networking solutions are explored in the next section.

Despite its benefits, the continuous-presence solution cannot on its own support all the IDL needs of K–12 schools. Data-oriented solutions, such as groupware and the Internet, should be implemented by K–12 schools to complement the continuous-presence solution.

2.5   NETWORKING SOLUTIONS

As identified in Chapter 1, distance learning networks need to cover the local and campus-level requirements, access requirements, metropolitan requirements, regional and national requirements, and global-level requirements.

A number of networking solutions are available to K–12 schools in supporting application solutions. Each solution differs in terms of the number of application solutions it supports and in terms of its ability to support the particular IDL needs of the students, the teachers, the parents, and the telecom/datacom administrators. At the access level, these networking solutions can be categorized into the following groups:

- Cable TV solutions;
- ATM-based solutions;
- Video dialtone solutions, such as ADSL, HFC, and FTTB;
- ISDN solutions.

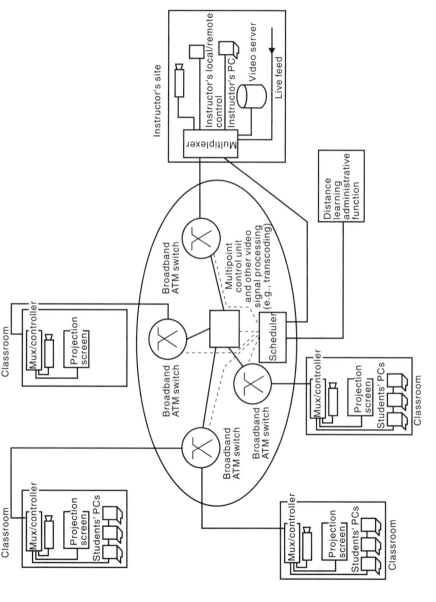

Figure 2.2 A state-of-the-art continuous-presence distance learning ATM-based network.

Each group of solutions is examined next.

2.5.1   Cable TV Solutions

The cable TV industry has been (and continues to be) active in supporting the application solutions described in the preceding section. Cable TV solutions can support the following IDL applications solutions:

- Access to the Internet;
- Groupware communications;
- Educational TV programming;
- One-way video/two-way audio solutions.

In addition, cable TV providers are currently conducting trials of two-way video/audio solutions. Perhaps the most important service that cable TV companies can offer to K–12 schools is the emerging data-over-cable solution.

Data-over-cable networks are offered by cable TV network providers to support the needs of local communities in suburban areas, including residential customers, city halls, court houses, and libraries. Cable TV companies offer this service to take advantage of the extensive installed base of coaxial cable, which reaches a high percentage of high schools and private residences [14].

Figure 2.3 shows a generic architecture of a data over cable network. The data over cable system includes three main elements: a bridge or router, a frequency translator in the headend, and a cable modem. The router or bridge provides K–12 students with access to application solutions, such as Internet access. The frequency translator located in the cable company headend converts the digital signal arriving from the router and converts signal to an analog channel. Multiple frequency translators can support multiple LANs. The analog signal is received by the cable modem/PC adapter card located at the schools uses subscriber data to modulate a 6 MHz carrier upstream.

Several manufacturers offer cable modems, including the following early entrants:

- *DEC.* The model name of DEC's cable modem is ChannelWorks Bridge. It costs $6,000 and supports 10 Mbps full duplex.
- *LANcity.* The model name of LANcity's cable modem is LANcity router. It costs $6,995 and supports 10 Mbps full duplex.
- *Zenith.* The model name of Zenith's cable modem is ChannelMizer. It costs $2,000 and supports 4 Mbps full duplex.
- *Intel.* The model name of Intel's cable modem is CableLink. It costs $6,000 and supports 10 Mbps full duplex.

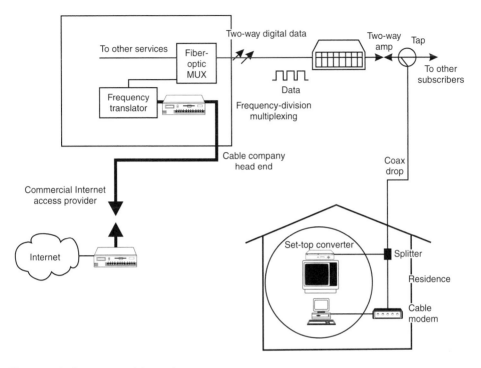

Figure 2.3 Data-over-cable services.

Several cable TV companies support data-over-cable TV services. For example, at press time, Continental Cablevision in Cambridge, Massachusetts, provided cable access to PSInet, a commercial Internet provider. Another example is TCI's data-over-cable TV services. These services support a community school network and Internet access in Glenview, Illinois. They also support IDL with the Utah Business Education partnership in Provo.

Data-over-cable solutions provide K–12 schools with the following benefits:

- *Support for data solutions.*
- *Adequate bandwidth.* The data-over-cable TV solution provides K–12 schools with high bandwidth that may reach 10 Mbps.
- *Availability.* Data-over-cable solutions can be easily made available to K–12 schools since more than 50% of schools are wired with cable TV. The total percentage of schools with cable drops range from 45% to a high of 75% [9].

Data-over-cable solutions do, however, have two major problems:

- *Reliability.* Data-over-cable solutions are not completely reliable yet.
- *Scalability.* Data-over-cable solutions are not scalable yet. If interconnection plans of high schools are limited to one district, then this solution is inadequate.

2.5.2 ATM-Based Videoconferencing Solutions

ATM-based solutions are emerging to address the data and videoconferencing needs of K–12 students. As discussed in Chapter 1, ATM refers to a high-bandwidth (45 Mbps to 622 Mbps now and more in the future), low-delay switching, and multiplexing technology that is now becoming available for both public and private networks. ATM principles and ATM-based platforms form the foundation for a variety of high-speed digital communication services, including LAN interconnection, imaging, and multimedia applications. ATM supports both switched (SVC) and nonswitched (PVC) connections. ATM can be used to support both connection-oriented and connectionless services (e.g., SMDS). ATM also supports emerging services, such as CRS and LAN emulation. The initial CRS will be PVC-based. This service offers users a high-performance service that can support voice, data, video, and multimedia applications at DS1, DS3, and SONET access rates [10].

ATM can support *all* the IDL application solutions described so far, including:

- Access to the Internet;
- Groupware communications;
- Educational TV programming;
- One-way video/two-way video solution;
- Continuous presence.

ATM can offer K–12 schools the benefit of high transmission speeds and provide corporations with the versatility and flexibility of integrated data, voice, video, and multimedia communications. A variety of vendors have been introducing ATM-based products, and a number of carriers (e.g., TCG) either already provide services or are poised to do so in the immediate future.

A recent development in ATM is the support of *available bit rate* (ABR) services, which will give carriers the ability to increase network utilization while facilitating new ATM service pricing. ABR services can be priced significantly less than other ATM classes of service, such as continuous and *variable bit rate* (VBR) services. ABR is ideal for TCP/IP traffic. It relies on closed-loop congestion management. As such, ABR is the best class of ATM service to cost-effectively transmit TCP traffic over ATM without significant cell loss. This is particularly useful for distance learning programs that entail transmission of large amounts of LAN/server-resident data.

Some observers view ABR as the most significant development in the world of ATM standards, since users receive cost-effective service and carriers can increase the trunk utilization of their networks up to 95% by selling bandwidth that remains unused. Many believe that ABR services lend themselves better to traditional LAN-to-LAN traffic than earlier standards that involve traffic contracts with the customer and policing by the carrier. Because LAN workstations are allowed to transmit at any time, it is difficult for LAN equipment to comply with a traffic contract. An ABR standard and a service based on that standard allows ATM networks to carry LAN-to-LAN traffic more efficiently and with better performance. By contrast, *unspecified bit rate* (UBR) has no guarantees; if the ATM switch in the network runs out of capacity, it will drop user cells. ABR, therefore, is more elegant in that it uses feedback to the source to increase or decrease traffic input; requires hardware support. Some prestandard ABR implementations may appear in 1996, but it will be at least late 1997 before general equipment interworking will be supported.

Rate-based traffic management is required to provide ABR services. Recent versions of the ATM standards include rate-based, closed-loop traffic management in addition to the open-loop mechanisms specified in earlier versions. Table 2.1 summarizes the currently defined ATM traffic types, their traffic descriptors, and *quality of service* (QOS) guarantees; these services can be used to support a variety of distance learning applications, including digital video.

Specifically, *constant bit rate* (CBR) service provides a constant bit rate pipe for applications such as voice, circuit emulation, and continuous bit rate video (JPEG/MPEG). *VBR real time* (VBR RT) service transports variable-rate information and provides tight bounds on delay to support applications such as voice with silence removal and compressed packet video. *VBR nonreal time* (VBR NRT) transports variable-rate information to support applications such as transaction processing. UBR service provides a best-effort delivery while making no QOS guarantees. ABR service provides for feedback-controlled sharing of spare network bandwidth to support elastic applications such as TCP/IP, IPX/SPX, and APPN.

Some users have applications that do not have fixed or predictable bandwidth requirements. Such elastic applications want the following: (a) as much bandwidth as possible; (b) access to spare bandwidth as fast as possible; (c) QOS guarantees; and (d) inexpensive service. For CBR and VBR, the user must declare parameters (PCR, CDVT, SCR, BT) at connection setup. For many applications, it is impossible to predict with accuracy those parameters in advance. As a consequence, the user must pay for unused bandwidth, delay bursting until SCR conformance, or exceed the traffic contract and lose the QOS guarantee. For ABR, the user states the maximum bandwidth possibly required (which would default to the access rate) and optionally a minimum usable

Table 2.1
ATM Services

Service	Descriptors	Loss Guarantee	Delay Guarantee	Bandwidth	Feedback
CBR	PCR, CDVT	Yes	Yes	Yes	No
VBR RT	PCR, CDVT, SCR, BT	Yes	Yes	Yes	No
VBR NRT	PCR, CDVT, SCR, BT	Yes	Yes	Yes	No
UBR	Unspecified	No	No	No	No
ABR	PCR, CDVT, MCR	Yes	No	Yes	Yes

CBR = constant bit rate
PCR = peak cell rate
CDVT = cell delay variation tolerance
VBR = variable bit rate
RT   = real time
SCR = sustained cell rate
BT   = burst tolerance
NRT = non–real time
UBR = unspecified bit rate
ABR = available bit rate
CR   = minimum cell rate

bandwidth, which the network will guarantee in the event of congestion. The network feedback will dynamically adapt the rate offered to the user, based on available bandwidth. Users are able to reliably burst on demand, for extended periods, with a guaranteed cell loss ratio or a guaranteed MCR.

CBR traffic uses fixed bandwidth. Service providers must allocate fixed amounts of their network resources to provision these connections. VBR consumes variable amounts of real resources, but bandwidth still must be reserved to meet the QOS requirements. As Figure 2.4(a) indicates, network utilization with just CBR and VBR is limited, with wasted transmission capacity.

UBR service tries to fill these wasted transmission slots, as can be seen in Figure 2.4(b). However, UBR service makes no QOS guarantees: There is no guaranteed bandwidth, and during network congestion there may be high cell loss. Either users suffer with useful throughput (actual throughput considering retransmissions due to cell discard) or service providers overprovision their network to provide minimally acceptable performance.

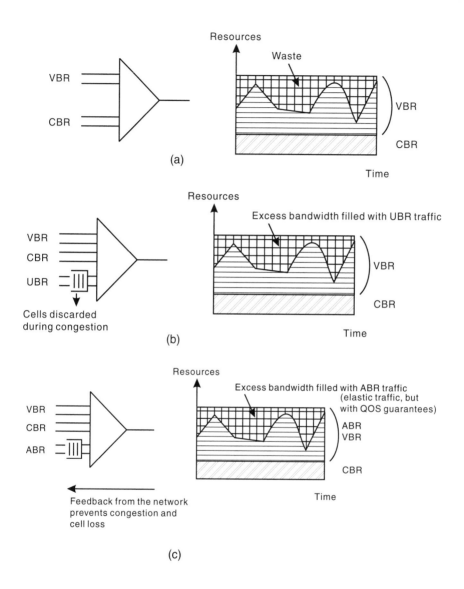

Figure 2.4 (a) Network carrying CBR and VBR traffic; (b) networks carrying VBR, CBR, and UBR traffic; (c) network supporting ABR; and (d) feedback mechanisms for ATM.

ABR is a new class of ATM services that permit users to dynamically access available bandwidth not being used by other ATM services, such as CBR and VBR services. Unlike CBR and VBR, which require users to predict and declare application bandwidth requirements and service providers to reserve

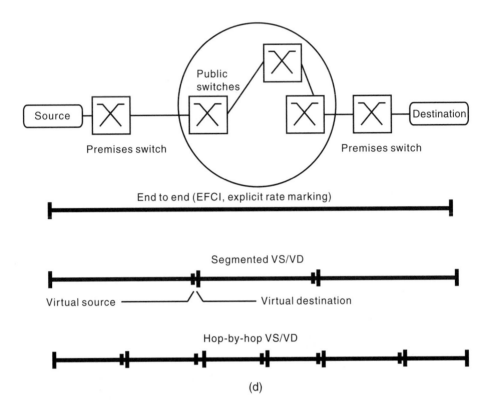

Figure 2.4 (continued)

fixed amounts of bandwidth to meet QOS requirements, ABR traffic is not po-
liced (or tagged "discard eligible"). ABR is dependent on sophisticated traffic
management mechanisms that monitor network traffic levels and detect the on-
set of congestion. Network feedback is used to adapt ABR traffic flows to the
available bandwidth in real time. That gives users the ability to reliably burst,
for extended periods of time, with minimal cell loss and maximum usable
throughput. Closed-loop feedback prevents congestion and adapts ABR traffic
to the available bandwidth in real time. Consequently, the service providers are
able to maximize utilization of the network, as can be seen in Figure 2.4(c).
Since excess bandwidth can be sold to users with a guaranteed QOS, the cost
basis for ABR is, in theory, marginal, compared to that of CBR and VBR services, which require bandwidth reservation.

Closed-loop feedback mechanisms include: (a) end-to-end explicit forward congestion indication (EFCI) with explicit rate marking; (b) segmented
virtual source/virtual destination (VS/VD); and (c) hop-by-hop VS/VD
(Figure 2.4(d) and Table 2.2).

Table 2.2
Feedback Approaches in ATM

Hop-by-hop VS/VD	Active hop-by-hop segmented
Segmented VS/VD	Active feedback control
	Segmentation (i.e., LAN/WAN, public/private)
Backward marking	Passive feedback participation
	Relies on end user
EFCI	Passive feedback participation
	Relies on end user

EFCI provides for only passive participation by the switch and represents the minimum required to comply with rate-based congestion management. EFCI relies on the end-user *network interface cards* (NICs) or equipment for compliance.

Backward rate marking provides an active means for the switch to participate in the congestion notification (by marking resource management cells), but it still relies on the user's device to resolve congestion.

Segmented VS/VD breaks the feedback loop into segments, at least one of which may be managed by the backbone service provider. This allows for better control of the scarcest resource, similar to a congestion "firewall": the firewall insulates the network from misbehaving users and allows for ABR service deployment utilizing existing non-ABR-ready end equipment.

Hop-by-hop VS/VD breaks the feedback loop into individual segments (one per pair of switches) and offers the tightest control, but it is also the most complex to build and manage.

An example of an ATM-based solution for IDL purposes is the well-publicized *North Carolina Information Highway* (NCIH). The network was created when the state of North Carolina asked BellSouth, Carolina Telephone, and GTE to develop a proposal for a high-performance network based on fiber optics, SONET, and ATM [11]. IDL was one of the first applications implemented on the NCIH. The origins of IDL preceded the NCIH and were part of the Vision Carolina IDL trial. Because of the success of the Vision Carolina trial, one of the objectives of NCIH was mapping this application to the NCIH and preserving the look and feel of the trial.

Figure 2.5 shows the Vision Carolina architecture. DS3 codecs in the classrooms were connected to a *multipoint control unit* (MCU) in the carrier's CO through DS-3 (45 Mbps) private line facilities. The MCU provides video bridging for multisite video delivery. The MCU consisted of a DS3 cross-con-

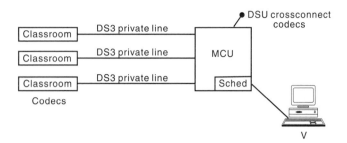

Figure 2.5   Vision Carolina architecture.

nect, DS3 codecs, quad-split devices, and a PC used for scheduling IDL sessions and for control during the sessions.

The mapping of the IDL application into the NCIH is shown in Figure 2.6 The key elements of the NCIH architecture for IDL are as follows:

- ATM switches of various carriers participating in the trial.
- SONET OC-3 (155 Mbps) or OC-12 (622 Mbps) lines interconnecting the ATM switches with each other. The switches are also connected to the classrooms through OC-3 lines.

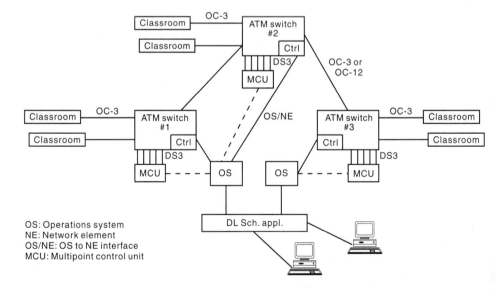

Figure 2.6   Current NCIH architecture for distance learning.

- Multiple operations systems (OSs) of multiple carriers, which enable the carriers to control the ATM PVCs on the switch. OSs are carrier computer applications used to run their business (e.g., order processing, service provisioning, troubleshooting).
- A distance learning scheduling application that, unlike the original Vision Carolina architecture, is not part of the MCU. Instead, the scheduling function is part of a separate application.

The process of initiating and conducting IDL sessions on the NCIH works as follows:

1. The IDL sessions are scheduled by communicating with a scheduling application.To schedule a session, the administrator has to identify the starting and ending times of the session.
2. The scheduling application reserves the resources required for the session, verifies that the classroom is available at the time, and assigns an MCU (directly attached to a specific switch) for the session. If all required resources are available, a confirmation message is sent to the scheduler of the session.

The NCIH supports a number of IDL applications, including access to museums, access to libraries, and remote education. The NCIH supports these applications through a number of services, including ATM CRS, SMDS, circuit emulation, and network control. The network control service enables users to schedule sessions in advance and for the network to establish connections automatically at the beginning of the session and to tear down the connection once the session is over. The multipoint bridging capability (that is, the MCU) enables multiple classrooms to be interconnected and to communicate via a videoconferencing session. This capability is provided though the MCU. For more information about the ATM architecture of the NCIH, the reader is referred to Chapter 6. Figure 2.7 depicts the U S WEST COMPASS distance learning ATM-based trial.

2.5.3 Asymmetric Digital Subscriber Line

As discussed, the access segment of a distance learning network is important because of the need to provide more bandwidth in the ramps to the broadband public network and because of the number of such links required (affecting total system cost). There are fiber-based, coaxial-based, and copper-based solutions to the access bottleneck. Fiber-based solutions are still expensive, and coaxial-solutions tend to be nondigital, thereby making them more difficult to support computers. This subsection looks at copper-based solutions, which provide one possible short-term compromise.

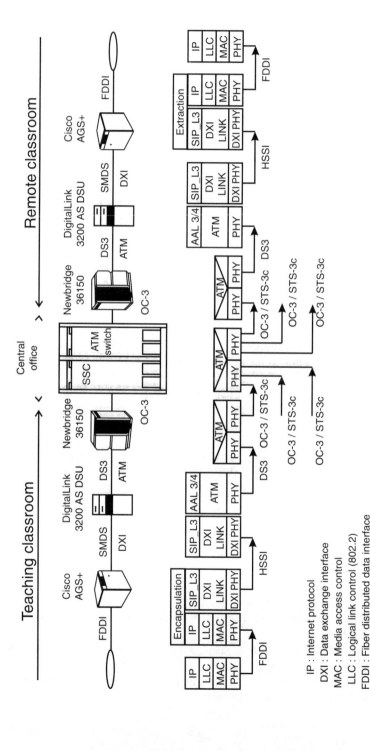

Figure 2.7 U S WEST COMPASS, HDDL trial: end-to-end protocol architecture for a data connection.

IP : Internet protocol
DXI : Data exchange interface
MAC : Media access control
LLC : Logical link control (802.2)
FDDI : Fiber distributed data interface
HSSI : High-speed serial interface

Until recently, only a few viable high-performance copper-based solutions ere available to the RBOCs to address the needs of local communities for oice, data, and video services. Recent advances in digital signal processing ave resulted in large increases in the amount of data that the telephone compa- ies can transmit over copper wire. These capabilities have been incorporated ı ADSL. As noted, ADSL is one of the merging VDT technologies that the BOCs are considering in meeting the broadband service needs of local com- unities, including K–12 schools, residences, and small businesses [6]. It is lso being considered for Internet access. ADSL has the following capabilities:

- One-way high bandwidth (downstream channel). From the CO to the user, ADSL may provide the user with data rates of 1.5 Mbps up to 6 Mbps. This bandwidth may be split among multiple channels. For example, 6 Mbps may be split among four channels, each capable of supporting a VCR-qual- ity video signal (MPEG-1 compression at 1.5 Mbps). This bandwidth may also be used for one channel, now capable of supporting fast-paced and sports-event quality (MPEG-2 compression at 6 Mbps) real-time video signal.
- One bidirectional ISDN "H0" channel operating at 384 Kbps;
- One bidirectional 160-Kbps ISDN basic rate channel containing two B channels, a D channel, and associated ISDN operations channels;
- A control channel, which enables K–12 schools and residential customers to control service delivery, including the ability to fast forward, reverse, search, and pause movies by relying on a remote control;
- An embedded operations channel for internal systems maintenance, audits, and surveillance;
- Passive coupling of ADSL to basic phone service. ADSL-based services are delivered to K–12 schools and other residential customers over the same copper pair that support basic telephone services. The coupling of ADSL and the basic phone line is passive; that means that if the ADSL system fails, the user can still place and receive phone calls.

ADSL can provide K–12 students with these capabilities by relying on *dis- crete multitone technology* (DMT). This technology maximizes throughput over a copper twisted-pair by dividing available bandwidth into 256 subchannels and allocating incoming data to each subchannel according to its ability to send data. If any subchannel cannot send data, it is shut off. The other working sub- channels are modulated so that they are able to carry from 1 to 11 bits per sym- bol based on learned channel characteristics. In a DMT-based ADSL system, a copper twisted-pair may be divided into three channels: (1) POTS occupies the baseband and is split from the digital data channels by passive filters or other means; (2) an upstream digital channel carries data in various configurations up to 384 Kbps; and (3) a downstream channel occupies the remaining bandwidth and carries up to 6 Mbps of asymmetric bandwidth as well as the downstrea portion of bidirectional channels (up to 576 Kbps).

ADSL is currently being evaluated by a number of RBOCs. For exampl Bell Atlantic has won a patent for its approach to ADSL. The patent describes system that provides digital video signal from a video information provider one or more subscriber homes. The subscriber orders programming by eith relying on a standard telephone set or by using a dedicated control device ov a packet link. Following a service order, the carrier establishes a link betwee the information provider and the CO serving the subscriber. Connectivity b tween the CO and the subscriber is established using ADSL. Interface units us frequency multiplexing technology to deliver digital video information wit voice information to the subscriber and support transmission of a reverse co trol channel. These interfaces allow baseband signaling and audio for conve tional telephone service.

ADSL can potentially offer K–12 schools several benefits, including (a) th ability to support simultaneous access to voice, data, and video services over single line (although video is only one way); (b) higher bandwidth than ISD BRI in one direction; and (c) the ability to receive voice services. Through thes capabilities, ADSL can support the following IDL applications: access to th Internet, groupware communications, and educational TV programming.

ADSL can also partially support the one-way video/two-way video solu tion links between schools sharing the same districts. It is unlikely, howeve that the RBOCs offering ADSL can establish these one-way video links wit universities. Note that only one program per ADSL link can be supported; tha means that a school with multiple classrooms needing simultaneous and dis tinct video would need multiple lines, which is unrealistic. Fiber would b much more suited for a school requiring more than one video feed. In addition ADSL may not be an available networking solution for several years, if ever Perhaps it will fare better as a way to provide more bandwidth to private resi dences, although it is also feasible that fiber will win that market, as discusse in Chapter 1.

2.5.4 ISDN

This interconnection approach involves the use of switched *digital* facilities t link a student with a training site located across the city or the country. ISDN provides end-to-end digital connectivity with access to voice and data service over the same digital transmission and switching facilities. It provides a rang of services using a limited set of user-network interface arrangements. ISDN supports narrowband and wideband solutions through three channels types B channels, D channels, and H channels. The *B channel* is a 64-Kbps acces channel that carries customer information, such as voice calls, circuit-switched data, or packet-switched data. The *D channel* is an access channel that carries control or signaling information and, optionally, packetized customer information; the D-channel has a capacity of 16 Kbps or 64 Kbps. The *H channel* is a 384-Kbps, 1.536-Mbps, or 1.920-Mbps (Europe) channel that carries user information, such as video teleconferencing, high-speed data, high-quality audio or sound program, and imaging information.

ISDN defines *physical user-network interfaces*. The better known of these interfaces is 2B+D, which provides two switched 64-Kbps channels, plus a 16-Kbps packet/signaling channel (144 Kbps total); other well-known interfaces are shown in Table 2.3. Other ISDN interfaces include 1B+D, which is currently under trial by some LECs (e.g., NYNEX), and NB+D, where N is greater than 2 but less than 23.

Table 2.3
ISDN Interfaces

23B + D	Twenty-three switched 64-Kbps channel plus a 64-Kbps packet/switching channel (1.536 Mbps total)
H0 + D	Switched aggregated 384-Kbps links
H10 + D	Switched aggregated 1.544-Mbps links

ISDN is increasingly becoming a networking solution that can support the following IDL applications:

- *Access to the Internet.* RBOCs are increasingly addressing the needs of the K–12 schools for access to the Internet through ISDN solutions.
- *Access to groupware.* ISDN can support both public and private groupware solutions.
- *Two-way videoconferencing solution.* The RBOCs are developing this capability to meet the needs of K–12 schools

ISDN provide K–12 schools with the following benefits:

- *Adequate bandwidth.* Many ISDN solutions are available that can support the bandwidth needs of K–12 schools for videoconferencing solutions.

- *Availability.* The RBOCs have been actively deploying ISDN to support the networking needs of K–12 schools. ISDN (2B+D) provides K–12 schools with a total bandwidth capability of 128 Kbps.

    ISDN in its basic form, however, cannot provide students with access to TV programming. That is the role the RBOCs have allocated to VDT services.

2.6 SUMMARY

IDL can reduce the isolation of rural communities by linking them to education providers, such as universities. IDL can also eliminate the electronic isolation of the traditional classroom by linking them with museums, libraries, universities. Five major distance application solutions were examined in this chapter. Table 2.4 correlates these solutions with the applications they can support. A range networking solutions are available to support these IDL applications solutions, as shown in Table 2.5

Table 2.4
K–12 Distance Learning Application Solutions

| Applications Solutions | *K-12 Applications* | | | | |
	Colaborative Learning	Access to Museums	Access to Libraries	Remote University Education	Remote Education Links to Other Schools
Internet	×	×	×	×	×
Groupware	×	×	×	×	×
Educational TV		×		×	
One-way video/ two-way video two-way audio		×	×	×	×
Two-way audio/video	×	×	×	×	×

Table 2.5
K–12 Distance Learning Networking Solutions

Applications Solutions	K-12 Applications				
	Cable TV	*ATM*	*Satellite*	*ISDN*	*ADSL*
Internet	×	×	×	×	×
Groupware	×	×	×	×	×
Educational TV		×	×	×	
One-way video/ two-way audio		×	×	×	×
Two-way audio/video	×	×	×*	×	×†

* Using two send-receive Earth stations.
† Much lower quality link on the return side.

References

[1] Portway, P. S., and C. Lane, *Technical Guide to Teleconferencing and Distance Learning*, Applied Business Telecommunications, 1992, p. 281.

[2] Portway, P. S., and C.Lane, *Technical Guide to Teleconferencing and Distance Learning*, Applied Business Telecommunications, 1992, p. 128.

[3] Foshee, D., "Two-Way Compressed Video in Education," *Teleconference*, Vol. 13, No.1, p. 33.

[4] Minoli, D., *Analyzing Outsourcing, Reengineering Information and Communication Systems*, New York: McGraw-Hill, 1995.

[5] Stecklow, S., "Internet Becomes Road More Traveled as E-Mail Users Discover No Usage Fees," *Wall Street Journal*, Sept. 2, 1993.

[6] Minoli, D.,*Video Dialtone Technology, Approaches, and Services: Digital Video over ADSL, HFC, FTTC, and ATM*, New York: McGraw-Hill, 1995.

[7] Kobielus, J., "Groupware Buyer's Guide—The Time Is Ripe to Pick Groupware," *Network World*, Aug. 9, 1993, p. 47.

[8] Van Tassle, P., "The Global Village in the Classroom," *New York Times*, Mar. 21, 1993, Section 13, p. 1.

[9]   Yankee Vision, Consumer Communications, "Education's Electronic Superhighways," *The Yankee Group*, Vol. 10, No. 6, May 1993, p. 20.

[10] Minoli, D., and M. Vitella, *Cell Relay Service and ATM for Corporate Environments*, New York: McGraw-Hill, 1994.

[11] Grovenstein, L. W., C. Pittman, J. H. Simpson, and D. R. Spears, "NCIH Services, Architecture, and Implementation," *IEEE Network*, Vol. 8, No. 6., Nov./Dec. 1994, pp. 20–21.

Bibliography

Hawkins, J., "Technology and the Organization of Schooling (Technology in Education)," *Communications of the ACM*, Vol. 36, No. 5, May 1993, p. 30 (6).

Layland, R., "A Gateway to Internet Health and Happiness," *Data Communications*, Sept. 21, 1994.

"Telco Seeks to Educate Congress on Cost of Connecting Schools," *Telco Business Report*, Vol.11, No. 2, Jan. 31, 1994, p. 1.

(Re)Educating the Corporation

3

Many U.S. corporations are now facing the challenge of competing effectively in the global economy. These corporations cannot meet that goal without fostering a rapid and continuous process of learning among their employees, partners, and suppliers. Learning, however, is more than just corporate training and university education. It also encompasses collaborative research, in which employees share learning with each other, with customers, and with suppliers. Given this perspective, the purpose of this chapter is to: (1) explore the learning problems facing U.S. corporations and (2) define the existing and potential roles of distance learning technologies (applications and networking) in solving those problems.

To accomplish this mandate, the chapter first identifies the distance learning challenges facing U.S. corporations. The chapter than identifies the key distance learning applications that can meet those challenges. The chapter then describes each application.

The chapter is targeted at individuals who undertake *corporate support functions*, including line managers, financial managers, and accounting managers. The chapter is also targeted at people undertaking primary corporate functions, including marketing, customer service, and operations. In addition, the chapter is targeted at network service providers seeking distance learning marketing opportunities in the distance learning industry.

3.1 CORPORATE DISTANCE LEARNING CHALLENGES

Corporate management faces a number of business challenges, including the following:

- Upgrading the skills of corporate employees through corporate training programs;

- Upgrading the skills of corporate employees through formal university education;
- Establishing closer electronic ties with customers;
- Developing products and services rapidly and responding quickly to customer needs;
- Improving communications between upper management and employees.

3.2 CORPORATE DISTANCE LEARNING APPLICATIONS

IDL solutions can play a significant role in addressing each of the issues identified in the introductory section. To address those challenges, telecom/datacom managers need to understand how the challenges translate into specific IDL communications systems, applications, and solutions. Figure 3.1 provides a conceptual framework of the electronic links that corporate information entities at headquarters need to establish with remote corporate entities, as well as with customers and suppliers. These links are necessary to support the following IDL applications:

- Corporate training;
- Intracompany collaborative learning and research;
- Intercompany collaborative learning and research;
- Remote university education;
- Executive communication;
- Market research.

3.2.1 Corporate Training

Domestic corporations face growing competition from Japanese and European companies. Global competition will further intensify in the coming years as emerging Southeast Asian countries join the developed countries in competing head-to-head with the United States. The challenge facing U.S. corporations is how to prepare their workforce for effective participation in the skill-intensive industries that increasingly represent the backbone of the U.S. economy. Examples of these industries include computing, biochemical products, and telecommunications. The United States can no longer compete in low-wage, labor-intensive industries. Those industries have moved to the Far East and to third-world countries. Skill-intensive industries are increasingly replacing traditional labor-intensive industries as the key creators of new jobs in the U.S. economy. The success of U.S. corporations in these new industries will depend to a greater extent on the relative skills of their workforce and to a lesser extent on the availability of natural resources and the cost of labor.

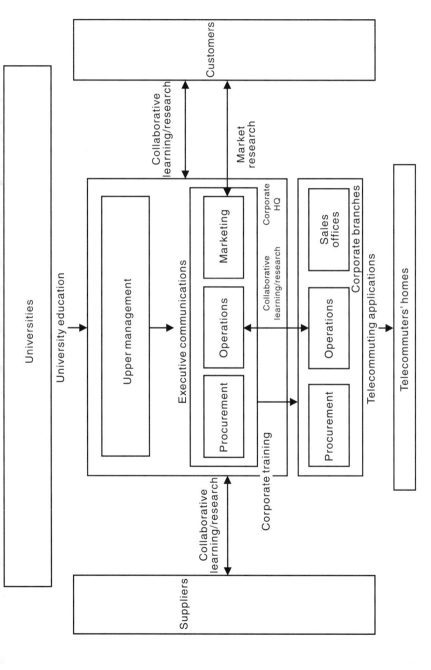

Figure 3.1 Corporate distance learning links and applications.

3.2.1.1   Corporate Training Participants

As Figure 3.1 shows, corporate training applies to corporate functions located in field offices and other remote corporate locations. This subsection describes the IDL needs of each of the remote corporate functions.

Blue-Collar Workers

One group of corporate employees that requires intensive training is blue-collar workers. Their training is necessary to increase productivity and to enhance their abilities to manage increasingly complex machines. According to experts in the field of corporate training, over 50 million U.S. workers need to be retrained. Workers need to acquire basic skills, including reading, writing, and computation, as well as interpersonal skills. Interpersonal skills are necessary because workers can no longer count on maintaining their jobs by performing repetitive jobs that are defined by rigid work rules. Instead, workers need to learn how to be more adaptive to be able to perform multiple tasks in a flexible manufacturing environment. They also need to learn how to operate as part of a team, since the teaming model is increasingly replacing the older model, in which workers were assigned specific tasks that they performed repeatedly. The previous paradigm limited the need for worker cooperation and multiple skills.

White-Collar Employees

While (re)training blue-collar workers is in itself a major national task, it is not the only demanding training task faced by U.S. corporations. The challenge of corporate (re)training also encompasses white-collar workers, who have to keep up with the latest developments in their fields. Technical functions, which represent about 80% of the U.S. workforce and receive about 30% of the annual corporate training budget, have the most demanding training needs. For example, technical applications are advancing so rapidly that a person performing a technical function needs to be constantly retrained to keep up with the latest advances in the field.

Product training is another important area of training for white-collar workers. In a competitive economy, corporations are racing to introduce new products and services. To be more effective, they need to speed the process and reduce the cost of disseminating information about new products to the sales force in the field. Such dissemination ensures that the sales force is educated about the specific features of new products and the value those products create for corporate customers. To accomplish this, organizations are seeking cost-effective IDL alternatives to the traditional face-to-face sales meeting paradigm. The need for such alternatives becomes more critical as (a) the size of the sales force increases; (b) the sales force is more distributed among multiple locations, which may be scattered throughout a state, the country, or the world; and (c) the rate of new-product introduction increases. The alternative to direct face-to-face meetings could be one of several IDL technologies. Marketing and product managers play a vital role in the development and introduction of these products to salespeople in the field.

3.2.1.2 Problems With Current Corporate Training

Despite the importance of training, the United States has the lowest percentage of spending on training compared to the advanced European nations and Japan. The United States currently spends approximately $30 billion on corporate training, and about 10% of the U.S. workforce are beneficiaries of that training. While the monetary amount appears large, it pales in comparison to U.S. corporate expenditures on retooling (new plants and equipment), which exceed $475 billion. Considering the growing importance of human capital relative to the capital allocated to retooling, corporations need to find a better balance between the two outlets of capital expenditures. (Obviously, government can play a role in encouraging corporations to invest in education; this role of government is explored later.)

3.2.1.3 Technology Needs of Corporate Training

The technology requirements associated with corporate training differ from one function to another.

- Technical personnel are the most demanding in terms of the technology they need for corporate training. They value high-quality two-way video-conferencing solutions that can transmit quality video. Often they also need high-resolution still images related to the topic they are studying, which can range from CAD images for engineers to medical images for health services personnel.
- Marketing personnel may also be demanding in terms of the technology they need for product training. The quality of service required for product training is a function of the complexity of the product being introduced. If the topic of training is, for example, a new insurance product, the IDL solution may be as simple as audio conferencing. As the products become complex (e.g., technically oriented products), marketing managers need to add the visual component to their conferencing systems.
- Administrative personnel, such as accountants, lawyers, and human resource managers, are less demanding in terms of their IDL technology requirements and may be satisfied with one-way video/one-way audio solutions.

3.2.2 Executive Communication Applications

Corporate management in a headquarters location needs to establish (distance learning) communications links with corporate employees located in branch offices to formulate, create and disseminate corporate policy and to support emergency communications.

Corporate Policy Communication

The financial success of corporations is increasingly dependent on the ability of the upper management to anticipate competitive moves, to rapidly formulate and implement effective responses to those moves, and to mobilize employees to assume responsibility in implementing those moves.

To ensure the effective dissemination to and buy-in of employees, management needs to establish a continuing dialog with as many employees as possible throughout the process of developing and implementing the company's strategic plans. That dialog would accomplish two purposes: (1) to generate feedback from employees about planned changes, and (2) to create consensus among employees to ensure that the strategies are implemented, not undermined because of the lack of employee support or understanding. The keys to the effectiveness of these interactions are *speed, cost, interactivity,* and *reach.* Reach refers to the number of employees that can be contacted simultaneously by corporate management; interactivity is the ability of management to establish a two-way dialogue.

Traditionally, managers have communicated their messages and requested feedback from employees through memoranda or through field visits to branch offices. Increasingly, managers rely on e-mail to reach employees in various locations, an approach that enables employees to respond to the messages. E-mail is being complemented by other telecommunications approaches, including video broadcasting.

Emergency Communications

Corporations have been increasingly faced with emergencies, such as the Tylenol-tampering scare and the 1994 Intel defective-chip affair. Corporate management needs to rapidly reach employees in case of emergencies to ensure that all employees are making a concerted effort to contain and diffuse a crisis with minimum damage to corporate profits and corporate reputation. Upper management is (or should be) seeking cost-effective IDL approaches. The type of conferencing technology may be a function of the products being sold.

3.2.3 Intracompany Collaborative Learning and Research

In a competitive economy, corporations are racing to introduce new products and services. Product introduction involves extensive interactions at various stages of product development among engineers, product designers, sales people, marketing researchers and managers, and R&D. A successful process of idea generation, product development, and implementation must be *rapid, interactive, frequent*, and *inclusive* of all the departments and functions involved. It also must be a learning process in which all participants learn from each other and build on each other's knowledge and ideas. Of all the applications that have been explored in this chapter, this IDL application is probably the most demanding because it requires not only the visual/audio component of a conferencing solution but also the data component.

3.2.4 Remote University Education

Corporations provide donations and R&D-related funds to universities in exchange for R&D cooperation, recruitment assistance, and educational services. Educational services include university degrees offered to employees during after-work hours, as well as continuing education.

The corporate approach of meeting employees' needs for university courses and degree has been for employees to attend classes on campus. One shortcoming of this traditional approach is that many employees are unable to attend courses on university campuses due to physical distance.

3.2.5 The Telecommuting Distance Learning Application

Telecommuting is a growing corporate phenomenon that enables employees to work at home several times a week while retaining electronic communications links with the corporate office. According to industry figures, between 4 million and 8 million corporate U.S. employees have worked at least 8 hours per week from home as company employees and during business hours in recent years [1]. Employees are increasingly requesting permission to telecommute as a way to improve productivity, and at the same time spend more time with their families by reducing time on the road. Telecommuting saves money that employees spend on purchasing, maintaining, and fueling a car, on public transportation, and on summer and winter clothes. Employees believe that as telecommuters they can be more productive because they can work with no interruptions from coworkers, in more comfortable surroundings, with reduced exposure to air pollution, particularly in major urban areas, and without the stress associated with commuting.

Faced with increasing U.S. and international competition, regulatory pressures from federal, state, and local government, and increasing demands from their employees, employers are now developing and implementing telecommuting programs. Productivity gains can be achieved because telecommuters can work longer hours (6 to 10 hours per week) in view of eliminated commuting. Telecommuters also have improved work quality due to easier management of distractions and stronger work incentives resulting from a greater sense of being in control.

Employers are also developing and implementing telecommuting programs to reduce the costs of recruiting, relocation, and health insurance; to lower overhead costs associated with office space by making double use of desks; to reduce retraining costs resulting from employee turnover; and to recover from disasters more quickly by establishing communications links between corporate offices, including corporate data center locations, and the telecommuters' homes. Other sources of cost savings to employers include reduction in travel costs and reduction in penalties associated with non-compliance with government air quality mandates, which translates into fewer costs associated with penalty fees. Telecommuting also enables employers to reduce costs by utilizing untapped resources, including over 30 million disabled individuals who could join the workforce through solutions such as telecommuting.

The needs of telecommuters related to IDL are similar to those of their corporate colleagues working out of corporate offices. These needs include the following:

- *Collaborative research and education* with their corporate colleagues operating at corporate offices. Telecommuters would like to establish computer conferencing sessions with supervisors, partners, and technical colleagues on short notice to collaborate in the process of developing new products. For example, team members residing in different locations may be working on a design problem; establishment of a voice and data conference session would allow all the participants to view product illustrations, specifications, and graphics on their computer screens. That way, problems are discussed and resolved in real time. At the end of the conference, the latest version of the product design is available to all the participants and is saved in individual PC files or in a LAN server.
- *Access to university courses from their residences.*

3.2.6 Intercompany Collaborative Learning and Research

To speed the process of product development, corporate researchers, engineers, and marketing managers need to establish collaboration and learning links with their counterparts in other corporations, including their vendors and suppliers.

A successful process of idea generation, product development, and implementation must be *rapid, interactive, frequent*, and *inclusive.* Such a work paradigm can be viewed as another demanding IDL application. This application requires not only the visual/audio component of a conferencing-based solution but also the data component.

3.2.7 Market Research

To compete more effectively, marketing managers need to develop product features that meet customer needs. To keep in touch with their customers, many marketing managers conduct primary research, that is, gather information directly from customers. Traditionally, primary research has involved face-to-face meetings, telephone interviews, survey questionnaires, and focus groups. Each of these approaches has its limitations. For important customers, corporations may want to use complementary videoconferencing, which can be considered a distance learning application.

3.3 THE NEEDS OF IS CORPORATE MANAGERS

In selecting corporate IDL solutions, telecom/datacom managers should take into account the following factors:

- *Geographic coverage.* Telecom/datacom managers seek application solutions that support the geographic range of their end users. This range may encompass LANs, MANs, and WANs.
- *Range of applications and functions supported.* Telecom/datacom managers are concerned about a larger set of applications than those that are important to individual IDL receivers and providers. By definition, the needs of the entire organization have to be taken into account.
- *Support for installed base.* In selecting a distance learning solution, telecom/datacom managers seek a distance learning solution that enables them to overlay the IDL solution on their existing network infrastructure. The corporate infrastructure may include the following elements:
  - *A voice networking environment*, supported by a PBX or centrex and extended over MANs or WANs through either private lines or a public switched network system. Private lines for voice applications are increasingly being replaced by virtual private solutions and by circuit switched services.
  - *Legacy systems, including SNA. System network architecture* (SNA) systems interconnect hosts in one location (usually the headquarters location) with terminals in the same location or in other locations. Traditionally, hosts and terminals were interconnected through X.25

networks and private lines. Increasingly, these locations are interconnected through a frame relay network.

- *A client/server environment.* Clients and servers may be in the same location and interconnected across LANs, such as Ethernet, token ring, and *fiber distributed data interface* (FDDI), or they may be remotely located. Until recently, client/server environments were extended over WANs through private line services. Now, FRS, ATM, and NMLIS are replacing private lines as WANs of choice to interconnect these environments.
- *Satellite networks.* This technology is used widely for high-quality video distribution.
- *Access methods and speeds.* Telecom/datacom managers need different access methods and speeds to support these applications. Their requirements are shaped by the installed base as well as by cost factors. The range of access methods and speeds may differ from one company to another.
- *Ease of use and administration.* Telecom/datacom managers value application solutions that are easy for IDL receivers and providers and other corporate end users to use.
- *Low cost.* Telecom/datacom managers should take into account a number of elements in measuring the total cost associated with a new application solution, including the cost of acquisition, implementation, maintenance, and training.
- *Intracompany versus intercompany communications.* The appropriateness of the application solution may depend on the extent of the needs of telecom/datacom managers for intercompany versus intracompany communications.

3.4 DISTANCE LEARNING APPLICATION SOLUTIONS

The U.S. corporate training market is estimated to be a $100-billion-a-year business; upward of 35 million individuals receive formal, employer-sponsored education each year. A fair portion of that market can be served by wide range of IDL applications solutions which include the following:

- Dedicated institutional IDL systems, which can be further classified, in increasing order of complexity and sophistication, as one-way video/two-way audio, two-way video/two-way audio, and n-way video/n-way audio with continuous presence;
- Groupware;
- The Internet.

The following descriptions of these solutions assess to what extent each solution can support IDL applications.

3.4.1 The One-Way Video/Two-Way Audio Solution

This IDL solution is also referred to as "business TV." Business TV is based predominantly on satellite technology. As discussed in Chapter 1, two satellite approaches are available:

- Traditional satellite links, in which a transponder is dedicated to video;
- VSAT technology, which is less expensive than traditional satellite approaches but more limited.

Table 3.1 defines some key video-related terms that come into play in the discussion of production-level video.

Table 3.1
Key Video-Related Terms

Component video	Separate luminance and color signals supporting high-quality pictures.
Composite video	A video stream that combines red, green, blue, and synchronization signals into one so that it requires only one connector and/or connection. Composite video is employed by most television systems (e.g., NTSC, PAL) and VCRs.
Compression	Reduction in the number of bits used to represent an item of data.
D1 digital video	A high-end digital component video format that ensures minimal signal and generational loss.
Encoding (compression)	The process of compressing a digital and/or analog signal into a compressed digital signal (e.g., with MPEG-2 techniques).
Encoding (process)	A process, not specified in MPEG-1/MPEG-2, that reads a stream of input pictures or audio samples and produces a valid coded bit stream as defined in MPEG-1/MPEG-2.
Motion Picture Expert Group 1/2 (MPEG-1/2)	A set of international standards (ISO/IEC 11172 and 13818, respectively) for the digital compression of video to the 1.5- and 6-Mbps rate.
Noninterlaced video	Computer monitors use noninterlaced video, in which each line is scanned sequentially rather than in an alternative manner.
National Television System Committee (NTSC)	Organization that has set the TV standards for North America. Specifically for traditional TV, NTSC specifies a 525-line 30-frames-per-second format.

Table 3.1 (continued)

Phase alternate line (PAL)	Used in Europe. It employs 625 lines per frame and displays 25 frames per second (which results in more flicker than NTSC standards).
RGB	For red, green, and blue, the three additive colors used for television and computer monitor signals.
S-video	A video signal that employs two channels: luminance (Y; namely, brightness) and chrominance (C; namely, color). It is referred to as Y/C. Falls between component and composite video in terms of quality.
Sequential color and memory (SECAM)	TV system used in France and Russia (the chrominance is frequency modulated).
YUV	The color space used by PAL and some NTSC formats (Y is the luminance and UV are the color components)

3.4.1.1   Traditional Satellite Services

Commercial satellite communications began in the mid-1960s. Satellites now carry voice, data, and video traffic. Satellite communications offer certain advantages, such as broadcast capabilities and mobility. The technology, however, is under pressure from other high-quality transmission media that support broadband communications, particularly fiber optics.

Communications satellites act as relay stations in space for radio, telephone, and broadcast communications. Commercial communications satellites lie in the geostationary (geosynchronous) orbit at approximately 36,000 km (22,320 miles) over the equator. Small variations in the mass and the shape of the Earth affect a satellite's orbit, requiring that it be "repositioned" under spacecraft power to regain the proper position. More than 120 communications satellites occupy the 165,000-mile circumference of the geosynchronous orbit.

The characteristics of geosynchronous satellites include the following:

- Broad Earth coverage (corresponding to approximately one-third of the Earth);
- Orbital period of 24 hours;
- Round-trip delay of approximately 500 ms for half-duplex communication.

The location of a satellite is nominally defined by the longitude (degrees west) of the point on the Earth's equator over which the geostationary satellite appears to be positioned. Most satellite antennas must point to a predetermined area of coverage throughout the entire life of the satellite. The method used to control the antenna depends on whether the satellite is spin stabilized or three-axis stabilized. Spin-stabillized satellites are stabilized by the induced rotation of the satellite according to the principle of the gyroscopic effect. The antenna is mounted on a counter-rotated platform so that it appears to be stationary in reference to the Earth's surface. In a three-axis-stabilized satellite, the antenna is mounted on a limited-motion gimbal, giving it flexibility in pointing; this antenna system requires a more complex control system than the spin-stabilized system. However, three-axis stabilized satellites have larger solar cell arrays compared to the surface of a spin-stabilized satellite, allowing them to have more operational power.

Satellites that are not in the geosynchronous orbit require Earth stations with movable antennas to track them; such is not the case for geosynchronous satellites. In the United States, which is the largest user of satellite resources, several organizations provide a spectrum of communications services over geostationary satellites operating around 4 and 6 GHz and around 12 and 14 GHz. Additional satellites in this frequency band are precluded by the lack of suitable orbit slots; hence, consideration is being given to higher-frequency operation (around 20 to 30 GHz).

The advantages of satellites include the following:

- The superiority of satellites for point-to-multipoint transmission makes them ideal for geographically dispersed broadcast TV (which requires transmission from a single location to many affiliate stations or to satellite dishes on roofs) and for private networks (e.g., teleconferencing and communications between corporate headquarters and branch locations).
- Satellite antennas are relatively easy to install and are even mobile, when installed on a small truck, and enable any ground station to become a network node. That makes satellites ideal for reaching remote or thinly populated areas where fiber optics cables would not be economically feasible. Fiber optics systems also require the establishment of a right-of-way for the cable to be laid, which can be difficult and costly.
- Satellites can be easily reconfigured while in orbit to cover different geographical areas.
- Total network failures are unlikely with satellite systems, except for catastrophic failure of the satellite itself. Storm damage to individual antennas will not affect the rest of the network.

Some factors that affect the applications of the technology to business communications include orbital spacing (e.g., only about 60 degrees of total equatorial orbit space are suitable for domestic satellite use), Earth station cost, antenna size, transponder power, and security. Difficulties in the late 1980s in delivering payloads into space have affected the availability of capacity for a number of years. Additionally, satellite orbits are close to being fully utilized. Other limitations include a lack of intrinsic security and propagation delays.

Nonetheless, many communication applications have emerged, including distance learning. What follows is a partial list of companies that use full-motion TV satellite–based technology for IDL. (Chapter 11 describes a case study that utilizes satellite technology.)

- Aetna Life Insurance
- Alabama Power
- Amdahl Corporation
- American Express Travel Related Services
- American Trucking Association
- Bell Communications Research
- BellSouth Corporation
- Channel Home Centers
- Computerland Corporation
- The Deparment of Energy
- The Department of Housing and Urban Development
- Digital Equipment Corp.
- Edward D. Jones & Corp.
- Electronic Data Systems Corporation (EDS)
- Equitable Life Assurance
- Federal Express Corporation
- The First Boston Corporation
- Ford Motor Company
- GE Appliances
- GE Capital
- GE Medical Systems
- GE Supply
- Georgia Power Company
- Georgia PublicTelecommunications Commission
- General Motors Corporation
- GMI Engineering and Management Institute
- Gulf Power Company
- Healthcare Satellite Broadcasting
- Hewlett-Packard Company
- IBM Corporation
- IDS Financial Services
- John Hancock Financial Services
- Johnson Controls
- Kidder, Peabody & Company, Inc.
- MCI Telecommunications Corporation
- Merck & Company
- Michigan Information Technology Network
- Mississippi Power Company
- MultiMedia Marketing Networks
- NBC Cable
- New York Life Insurance Company
- Phillips Consumer Electronics and Whittle Communications
- Prudential Insurance
- Redgate Communications Corporation
- Savannah Electric and Power Company
- Southern Company Services
- Southern Nuclear
- The State of Georgia
- Unisys Corporation
- The Upjohn Company
- U S WEST Communications
- Whirlpool Corporation
- Xerox Corporation

3.4.1.2 VSAT Technology

VSAT integrates transmission and switching functions to support on-demand links for point-to-point, point-to-multipoint applications. VSAT systems were deployed in relatively large numbers in the mid-1980s to early 1990s. Now with the plethora of new communications alternatives, it is not clear if the investment in hundreds of customer-owned satellite antennae is cost-justified with a reasonably short payback period.

One of the technological advances in the space segment has been increased transponder power, resulting in improved signal strength. This implies an acceptable signal quality even when smaller and less expensive Earth stations are used compared to traditional satellite systems. The term *very small aperture terminal* refers to the size of the antenna dish, which is usually 1.2m or 1.8m in diameter. The network topology is a star. Equipment prices have dropped to the point where a 1.2-m transmit/receive antenna now costs around $5,000.

During the past few years, there has been a general switch from C band to Ku band, providing even more signal directionality and increased effective iso-tropically radiated power. The FCC allocated the frequencies from 11.7 to 12.2 GHz and from 14.0 to 14.5 GHz—portions of what is commonly referred to as the Ku band—for primary use by fixed-satellite communications. With Ku band, some adjustment for rain attenuation can be made by using a stronger signal.

Users of VSAT networks can take advantage of the inherently broadcast nature of satellite communications. They can add or move sites within the footprint without concern for signal loss or increased monthly charges, since the signal is being continuously beamed across an entire area; however, moving does incur installation charges.

VSAT solutions are ideal for networks that have a star configuration, where a host computer site communicates with a number of geographically dispersed remote sites. As a rule, the outbound volume should be greater than the inbound volume. However, other configurations can be accommodated as well. VSAT applications are primarily for data and include point-of-sale, credit authorization, inventory control, and remote processing. In the early 1990s, video applications accounted for about 20% of the total VSAT traffic, and voice represented about 5%.

A VSAT network consists of three major elements:

- The master Earth station (MES);
- A number of remote VSAT Earth stations;
- A host computer site.

The MES, an intelligent node, is the communications hub for the rest of the network. Key to the successful use of small Earth stations is the star network topology, which enables the powerful MES to compensate for the relative weakness of the VSAT end of the channel. The large MES transmits a powerful signal to the satellite, so that the receiving VSATs can capture a high-quality signal. The transmit signal of the VSATs is relatively low powered, and the MES is needed to receive it, especially in a point-to-point network, where the VSATs are communicating with each other. The MES antenna is generally from 5m to about 9m in diameter. The MES must be designed specifically to support the type of equipment in use at the customer's host site. Most vendors' systems are designed to accommodate IBM SNA-based equipment, LANs, or both and are plug compatibles. The MES performs a variety of essential functions, including transponder monitoring and host interfacing. In addition to its antenna dish, the hub Earth station consists of RF/IF electronics, the network switching system, and a network management computer. Whether the MES for a network is shared among several organizations or is dedicated to one organization is a key question. The average network payback period for a network with a dedicated MES is estimated to be about five years, and it may be as short as one year in some cases. The cost of the hub—up to $1 million—places a fully private VSAT network out of the reach of most organizations. However, potential users will have access to full-service, shared-hub network offerings from satellite carriers and others.

The number of sites required to justify the choice of VSAT is variable. The vendors offering shared-hub network services claim there is no minimum number of sites (the emergence of shared-hub services offered by carriers is an important element in the size of the market). Whether the cost of a VSAT solution will be justifiable for a given organization is a complex decision and must finally be made on the basis of a detailed analysis.

The VSAT video solution offers several advantages to corporations:

- *Low monthly charges.* VSAT can help companies overcome the burden of local access charges. It can be cheaper than a regular transponder. To provide a sense of the possible costs involved in regular video/TV distribution, Table 3.2 lists "occasional service rates" in force at press time (for Vyvx's ImageNet service).
- *Access methods and speeds.* VSAT supports speeds up of DS1; hence, it supports H.261/H.262-video (i.e., videoconferencing quality), which is generally adequate for corporate users.
- *Range of applications supported.* VSAT networks can support the following IDL applications: corporate training, remote university education, and executive communication.
- *Support for installed base.* It requires minimum changes to corporate telecommunications infrastructure.

- *Geographic coverage.* VSAT can cover a wide geographic area, for example, major regions of the country or the entire country.
- *Ease of expansion.* Users can add a capacity to a network by requesting additional transponder bandwidth. Remote satellite antennae have to be installed. Generally, additional locations can be brought on line within a matter of days.
- *Intracompany versus intercompany communications.* VSAT technology is appropriate for both intercompany and intracompany communications.

Table 3.2
Occasional Service Rates

Point-to-Point Usage	*One-Way*	*Two-Way*	*Nonprime One-Way*
15 minutes	$140	$210	$95
30 minutes	$250	$375	$150
45 minutes	$360	$540	$205
1 hour	$440	$660	$265
More than one hour	Prorated 1-hour rate in 15-minute intervals (reservation needed)	Prorated 1-hour rate in 15-minute intervals (reservation needed)	Prorated 1-hour rate in 15-minute intervals (reservation needed)

Point to Multipoint Service	
Reservation 1 hour or less	Rates are established by adding the appropriate one-way point-to-point charge and $50 for each additional destination per hour
Reservation greater than 1 hour	Rates are established by adding the appropriate one-way point-to-point 1-hour charge and $50 for each additional destination per hour, prorated in 15-minute increments

The VSAT solution does have a number of limitations:

- *Inadequate support for some IDL applications,* including intracompany collaborative learning and research, intercompany collaborative learning and research, and market research. The VSAT solution cannot support these applications adequately because interactivity is limited to audio communications.

- *Increasing transponder cost.* The cost will likely be on the rise since many currently orbiting satellites are approaching the end of their useful lives and will not be replaced until the payload backlog is reduced.
- *Delay.* While VSAT is adequate for business TV, the half-second delay associated with VSAT transmission makes this technology less than optimal for two-way video/two-way audio communications with high interactivity.
- *Security.* Businesses need to encrypt the messages prior to transmission.

Example of a Business TV Implementation

A company that has implemented a VSAT system as an IDL solution is Ford Motor Company. A business TV system is used by Ford to train its engineers on the latest development in their fields. Recognizing that the shelf life of an engineering degree can be as little as 3 years, Ford allocates an annual budget of $240 million to retraining its engineers. Through its IDL program, Ford employees earn graduate credits and engineering degrees. The program was developed with Purdue University, the University of Maryland, and Wayne State University.

Ford also uses satellite technology in Europe to support the needs of its employees for IDL. The VSAT network of Ford supports 400 Ford dealers [2]. Through the satellite network, dealers are provided with updated information on products and services. The network also enables the dealers to provide direct feedback to senior management though live, on-the-air interactions. The network averages 2–3 hours of programming per week. According Ford, "There is no doubt that the satellite network has brought the company and its dealers closer together" [2]. The network enables Ford to establish instantaneous links with its employees and saves them the cost associated with planning national meetings and with producing and distributing reams of paper.

The computer industry is another major user of VSAT technologies. Examples of companies that use this technology for IDL purposes are Texas Instruments, Hewlett Packard, and IBM.

3.4.2   Two-Way Videoconferencing Systems

Terrestrially based video systems fall into two categories:

- Full-motion analog (6 MHz) or digital (45 Mbps) video;
- DS1 and $n \times 64$ Kbps videoconferencing systems.

Full-motion systems can be expensive and are used by only a few dozen companies. Videoconferencing systems (e.g., those using $n \times 64$ Kbps digital services) are currently used by more than 50% of all Fortune 1000 companies [4]. The largest users of videoconferencing are medical and educational institutions, with over 28% of the videoconferencing installed base, followed by wholesale retail with 24% [5]. Corporations rely on videoconferencing systems to support the corporate training. For example, in the retail industry, videoconferencing systems are used to broadcast information on merchandise and new products to dispersed store locations. The timely delivery of this information provides store managers with several benefits, including the ability to be more responsive to the marketplace, the ability to reduce store costs associated with carrying unwanted inventory, and the elimination of discount sales to dispose of obsolete merchandise.

In the automotive industry, videoconferencing is used to keep car dealers informed of price changes, incentive programs, spare parts availability, and maintenance procedures [6].

3.4.2.1 Benefits and Drawbacks of Two-Way Videoconferencing

The two-way videoconferencing solution to corporate distance learning offers several advantages to corporations:

- *Access methods and speeds.* A wide range of networking solutions can support the two-way video conferencing solution.
- *Range of applications supported.* The two-way videoconferencing solution can support *all* the IDL applications: corporate training, remote university education, executive communication, intracompany collaborative learning and research, intercompany collaborative learning and research, and market research. In the case of the collaborative research/learning application, desktop solutions are superior to room-to-room solutions.
- *Support for installed base.* This solution requires minimum changes to the corporate telecommunications infrastructure.
- *Geographic coverage.* This solution can cover a wide geographic area.
- *Intracompany versus intercompany communications.* This solution is appropriate for both intercompany and intracompany communications.

The two-way videoconferencing solution, however, has one major limitation: cost. The cost factors differ, depending on the networking solution selected, the number of participants, and whether the solution is implemented room to room or desk to desk. Multicasting is more difficult with a terrestrially based system than with a satellite-based system.

3.4.2.2 Example of a Two-Way Videoconferencing System

An example of a two-way videoconferencing network is that offered by U S WEST. The U S WEST Star network service is based on a fully interactive, fiber optic–based solution. Examples of companies that have taken advantage of this network are Scientific-Atlanta and Zenith Electronics Corporation.

These companies provide cable TV engineers located in Minnesota with training. With the help of Wadena Technical College, in northwest Minnesota, engineers meet over the Star network by traveling short distances to one of two videoconferencing sites, which are located in Wadena and Wilmar. Scientific-Atlanta and Zenith provide the instruction from the Minneapolis suburb of Anoka. Each engineer pays a $25 registration fee and in return receives training in audio and video technologies.

3.4.2.3 Example of a Two-Way Video Distribution Network

This section focuses, for illustrative purposes, on two high-end video distribution services currently offered by Vyvx, Inc., a subsidiary of William Companies [10]. The services are in support of network-quality broadcast video.

ImageNet™ is a nationwide digital component transmission service that allows major video and film production houses, TV network facilities, advertising agencies, and corporate communications centers in the United States to communicate interactively among high-end graphics and production equipment. Component video is transmitted in real time end to end.

AtlanticVision™ is a one-stop video broadcasting (one-way video/two-way audio) service between the United States and the United Kingdom. The service is offered by a recently formed alliance between BT, Teleglobe Canada, and Vyvx.

The customer can supply either an analog TV signal or a digitized component video at 45 Mbps rate. The pricing structure of Vyvx takes into account the following factors:

- Direction of transmission (one-way video versus two-way video);
- Quality of service (ranging from presidential to economy);
- Access type (digital versus analog);
- Domestic versus international.

ImageNet is a nationwide digital component transmission service that allows major video and film production houses, TV network facilities, advertising agencies and corporate communications centers in the United States to communicate interactively among high-end graphics and production equipment (see Figure 3.2). Vyvx offers both the *digital codec* for inter-point-of-presence digital transmission and the inter-point-of-presence *digital transmission facilities*.

There are two nearly lossless methods of digitizing NTSC, PAL, or SECAM TV signals: *digital component video* and *digital composite video*. Digital component video, also known as 4:2:2, is a time-multiplexed digital stream of three video signals: luminance (Y), Cr (R-Y), and Cb (B-Y). The 4:2:2 refers to the ratio of sampling rates for each component. This format is also often called

D-1, referring to the tape format associated with the digital component recording. NTSC digital component video encoding systems have utilized the following parameters:

- Luminance sampling frequency: 13.5 MHz;
- Sampling frequency for color differences: 6.75 MHz;
- Pixels: 858 by 525.

At the final stage, the word length for digital image delivery is usually
between 8 and 10 bits, but to maintain precision more bits may be utilized,
particularly in the early stages of off-line processing (e.g., 16 bits). The ITU-R
601 standard supports both the 525-line, 60 fields/second format, and the
625-line, 50 fields/second format. The other lossless encoding method is digital
composite video, known as 4fsc. This format also consists of three components:
Y, I, and Q. I and Q are quadrature modulated and summed to the Y component. The result is a single information stream sampled at four times the color
subcarrier rate. The term 4fsc refers to "4 times the frequency of the subcarrier."
That format is often called D-2, referring to the associated tape format.

ImageNet has the following features:

- It provides compressed digital component video transmitted at 45 Mbps
  in accordance with ITU-R 601 and ITU-R 723. ImageNet also allows up to

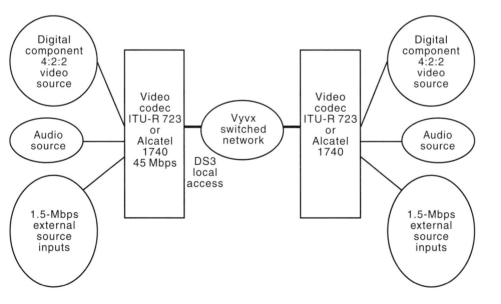

Figure 3.2 Vyvx service.

six 20-kHz AES/EBU digital audio channels transported at MPEG-1 layer 2 for high sound quality.

- ImageNet can support customer needs for either occasional service or dedicated service. The occasional service can be switched by customers to a variety of destinations at any time, anywhere in the United States.
- Vyvx can supply customers with digital component codecs and discount options.
- The service interfaces with standard video production equipment supporting ITU-R 601 digital video input standards.
- The service also supports international standard coding and transmission using equipment that conforms to ITU-R 723, and it supports Alcatel primary coding and transmission standards for their 1740 codec.
- Customers automatically receive 24-hour troubleshooting, network monitoring, and customer support.

Table 3.3 provides technical specifications for the services supported.

Table 3.3
ImageNet Technical Specifications for Component Video Transmission Service

Video coding specification	ITU-R 723
Audio coding specification	MPEG-1 LAYER II
Video I/F specification	ANSI/SMPTE 2J9M-1993 Serial component digital
Audio I/F specification	ANSI S4 40-1992 AES/EBV digital audio 4-6 channels
Transmission speed	DS3 44.734 Mbps
Data ports	DS1 1.554 Mbps
Access type	DS3 unchannelized

Table 3.4
ImageNet™ Technical Specifications for Alcatel 1740

Video coding specification	Alcatel 1740 VC
Audio coding specification	ITU-T J.41
Video I/F specification	SMPTE 12 SM-1992 Parallel component digital

Table 3.4 (continued)

Audio I/F specification	50–15 kHz, 600 Ohm, 4 Channel
Transmission speed	DS3 44.734 Mbps
Data ports	DS1 1.554 Mbps
Access type	DS3 unchannelized

AtlanticVision™ is a one-stop video broadcasting (one-way video/two-way audio) service between the United States and the United Kingdom. The service is offered by a recently formed alliance between BT, Teleglobe Canada, and Vyvx. The following are the key features of the service:

- Through the service, customers can send and receive broadcast-quality, full motion video via Teleglobe's CANTAT 3 transatlantic fiber-optic cable. The cable has a capacity of 96 digital fiber links and employs SDH technology.
- AtlanticVision™ transports signals between switching centers across North America to the main switching center in London (BT Tower).
- The service supports one-way digital transmission from either the United Kingdom or the United States.

In addition to ImageNet™ and Atlantic Vision™, Vyvx is currently developing and demonstrating new technologies and services, including digital archiving, ATM, remote movie production, and HDTV.

In the United States, access to Vyvx *television switching centers* (TSCs) is by way of analog facilities such as fiber, satellite, microwave, or coaxial cable. These services are called "first- and last-mile" service connections. The customer and the last-mile provider maintain responsibility for last-mile connections from the customer premises to the Vyvx *television center* (TC). However, Vyvx will provide special last-mile engineering services at customer's request.

In the service definition, TCs are the local access points to the Vyvx network with direct routing to the nearest Vyvx TSC. The TSC is the local-access point providing real-time rerouting, signal-quality monitoring, and service status and control.

Vyvx utilizes analog, rather than digital, first- and last-mile connections. Analog connections provide little degradation in picture quality while costing substantially less than digital connections. Additionally, Vyvx owns and maintains its codecs in its TSCs, further reducing the customer's transmission time.

In Canada, the first- and last-mile connections to TSCs are also established through analog facilities such as fiber, satellite, microwave, or coaxial cable.

However, the last-mile connections are provided and engineered by the Stentor-owner companies, which are BC Tel, AGT Ltd., Manitoba Telephone System, Bell Canada, NB Tel, Maritime Telephone and Telegraph, Island Tel, and Newfoundland Telephone.

Some establishments have "local accessibility." Those businesses, venues, TV facilities, cable systems, production companies, and other transmission networks are currently accessible to Vyvx through direct-loop or switchable connectivity at hubs. Vyvx/Stentor's accessibility to those local companies is considered "live," and connectivity can be accomplished soon after a request for transmission is submitted to Vyvx or Stentor.

A customer would use the following procedure to book a reservation from sites with existing connectivity:

1. Call Vyvx Traffic Center.
2. Supply event name, date of feed, start and end times of feed (EST), firm out or approximate end, and first- and last-mile city and circuit number.
3. Indicate if cross-connection at the hub will be made by the booking organization or by Vyvx.

The following procedure is used to book a reservation from a location without existing local loop or to request pricing information for installing local loops:

1. Check the Vyvx Local Connectivity Directory.
2. Call Vyvx TC.
3. Supply name of building or facility, street address and location of sites where local loops should be installed, local contact at facility, telephone number of contact, dates, and times.

Installing a local loop or getting pricing for the installation at most sites takes approximately a week to ten days, depending on the city and the available carriers; requests for sports facilities can sometimes be completed within eight hours, but it is best to allow at least 24 hours to order the loops from sports venues.

Although at a different range of service, Vyvx faces competition for business videoconferencing services offered by the major ICs, including AT&T, MCI, and Sprint. These ICs have a much larger share of the long-distance market (65%, 20%, and 13%, respectively).

Customers can choose from many sites in the United States and the United Kingdom, which are linked to the Vyvx network. These sites are summarized in Tables 3.5 and 3.6.

Table 3.5
First Video Network Affiliates

Albany	New York Network
Atlanta	Crawford Communications, Inc.
Baltimore/Owings Mills	Maryland Public Television
Boston	GBH Productions
Charlotte	Media Comm
Chicago	Nothwest Teleproductions
Cleveland	Classic Video
Dallas	AMS Productions
Denver	NORAC Production
Detroit	Producers Color Service, Inc.
Houston	Total Video, Inc.
Indianapolis	Sanders & Company
Jacksonville	Continental Cablevision of Jacksonville
Las Vegas	Creative Edge
Los Angeles	Pacific Television Center
Los Angeles	VDI
Miami	Comtel, Inc.
Minneapolis	Juntunen Video, Inc.
Nashille	TNN
Omaha	Cox Cable, Inc.
New York	All Mobil Video, Inc.
Newark	Prudential Television
Oklahoma City	Ackerman-McQueen, Inc.
Orlando	Century III
Phoenix	Southwest Television
Pittsburg	Production Masters, Inc.
Portland	KOPB
Raleigh/Durham	Capitol Satellite
St. Louis	Koplar Communications Center
Salt Lake City	STS Productions
San Antonio	Fibrcom
San Francisco/Oakland	Independent Television News

Table 3.5 (continued)

San Jose/Santa Clara	Transvideo Studios
Seattle	Third Avenue Productions
Southbend	Goldon Dome Productions
Tallahassee	Video Communications Southeast
Tampa	Telemation
Tulsa	Winner Communications
Washington	Interface Video Systems

Table 3.6
Canadian and U.K. Locations

Canadian Locations	*U.K. Locations*
Calgary	London
Edmonton	Belfast
Montreal	Birmingham
Ottawa	Bristol
Quebec	Cardiff
Regina	Carlisle
St. Johns	Granite Hill
Toronto	Kirk o' Shotts
Vancouver	Leeds
Winnipeg	Maidstone
	Manchester
	Newcastle
	Norwich
	Nottingham
	Plymouth
	Rowridge
	Southampton
	Tolsord Hill

The service is available as *occasional TV transmission service, cross-border occasional TV transmission service,* and *international service.* The following terminology applies to U.S. services (comparable terminology applies to international services):

- *Basic service—point to point.* Broadcast contribution quality TV transmission between two TSCs. Service is provided as one-way or two-way service in 15-minute increments.
- *Basic service—point to multipoint.* Broadcast contribution quality TV Transmission from one site to multiple sites. Service is provided as a one-way service.
- *Discounts.* Discounts are based on contract volume usage and are applied only to basic transmission services. At the end of the contracted discount term, charges are assessed based on the greater of the contractually committed hours multiplied by the applicable rate, or the actual hours used multiplied by the applicable rate. Applicable discount percentages are applied to the standard point-to-point and point-to-multipoint rates, thereby reducing the standard incremental rate for the appropriate reserved transmission period.
  - *Monthly usage discount.* Discount is based on committed monthly usage over a 1-month term.
  - *Annual usage discount.* Discount is based on committed annual usage over a 1-month term.
  - *Discount credit.* One-way service is credited 1 hour for each hour used. Two-way service is credited 1½ hours for each hour used. Multipoint service is credited 1 hour for each hour used.

Other charges include the following:

- *Overage.* Charges that are applied when a confirmed reservation is not canceled within a reasonable amount of time prior to the scheduled transmission start time (early acceptance, extensions, and overages do not apply if changes are made before the reservation begins).
- *Cancellation.* Charges that are applied when a confirmed reservation is not canceled within a reasonable amount of time prior to the scheduled transmission start time (early acceptance, extensions, and overages do not apply if changes are made before the reservation begins).
- *Rescheduling.* No charges apply for reservations that are rescheduled to a later time, provided the serving cities and duration of the reservation remain the same. Reservations that are rescheduled and then canceled are charged a cancellation charge of 100%. A reservation may be rescheduled only once (early acceptance, extensions, and overages do not constitute a rescheduling of a reservation).

- *Local switch charges.* Charges are applied when a customer requests that its first-mile or last-mile circuit be connected to another customer's first-mile circuit within the same TSC. Connections are for 24 hours.
- PGAD/secondary dropoff. A charge applies for each 1-hour increment of usage for each PGAD/secondary dropoff. PGAD/secondary dropoff originate from an authorized independent reservation but may terminate at different destinations.

Occasional-service rates for ImageNet were shown in Table 3.3.

3.4.3 The Internet

The Internet can support corporate needs for several IDL applications, including intracompany collaborative learning and research, intercompany collaborative learning and research, and corporate training.

The Internet supports corporate IDL applications through the following services:

- E-mail;
- File transfer;
- Host-to-host communications;
- Directory services;
- Online library catalog;
- Electronic whiteboards.

The Internet offers corporations the following IDL benefits:

- *Support by network service providers.* Access to the Internet is increasingly supported not only by the traditional Internet providers but also the AAPs, the LECs, and the IXCs.
- *A high-performance backbone.* The NSF backbone is migrating from a DS3 platform to a higher-performance network based on ATM and SONET.
- *An expanding range of applications.* The Internet can support collaborative learning and research applications (this was the original intent of the Internet).
- *Affordability.* The Internet remains one of the least costly approaches to providing interconnection.
- *Extensive information resources.* This represents another major source of strength of the Internet as an application solution to the distance learners. Through the Internet, the IDL community can access hundreds of libraries around the world, as well as library catalog and full-text delivery services.

- *Ease of use.* New services, such as WWW, make the Internet easier to access and to use, extending the use of the Internet beyond the corporate technical functions to encompass, for example, the marketing functions.

The Internet, however, cannot be considered as a complete IDL solution for the following reasons:

- *No guaranteed performance.* The IDL community needs to take into account that they cannot be guaranteed a given throughput across the Internet nor a consistent reliability level. That is because the Internet, while serving thousands of organizations and millions of individuals, lacks any mechanism for reserving bandwidth. In addition, the Internet is made up of many networks. Hence, the establishment of common reliability levels requires complex interactions among a large number of Internet providers.
- *Lack of extensive antiviral software.* Internet e-mail and downloaded files have been known to contain viruses. These viruses cannot be eliminated without the availability of antiviral software on every machine on the network.
- *Junk e-mail.* Corporations connected to the Internet can be flooded with useless and unwanted e-mail. Junk mail is particularly problematic to corporations, because their employees may spend valuable time reading the junk mail. Junk e-mail also consumes valuable disk space.
- *Security issues.* Probably the most important concern of the Internet is its security limitations. Several Internet security violations have occurred lately. The security limitation of the Internet restricts the usefulness of the Internet to interorganizational communications, including those associated with IDL solutions. For example, as long as security is a major concern, corporations cannot rely on the Internet to deliver corporate training courses, even when two-way videoconferencing over the Internet becomes more widely available.

3.4.4 Groupware

Groupware is an emerging data-oriented corporate IDL solution that can provide corporations with a private alternative to the Internet. As in the case of the Internet, groupware provides a solution to the collaborative learning application. While the Internet is a public solution, groupware can be implemented either as a private solution or as a public solution. Groupware refers to software that supports at least one of the following IDL applications:

- Electronic messaging;
- Data conferencing;
- Last messaging gateways.

Groupware offers IDL receivers and providers and other corporate employees several benefits:

- *Low cost.* In implementing groupware, telecom/datacom managers incur a nonrecurring cost. This cost is on either a user or a server basis.
- *Multiple applications.* Groupware supports multiple applications, including collaborative learning and research.
- *Ease of use.* Multifunctional groupware solutions are based on GUIs, which makes it easy to use by IDL receivers and providers and other corporate employers.
- *Security.* As a private solution, groupware provides a more secure solution for intracompany communications than the Internet.
- *Training and consulting.* The leading providers of multifunctional groupware provide their buyers with training or consulting as part of their product package.

While groupware offer corporations several benefits, telecom/datacom managers need to consider that groupware is predominantly implemented as a private application solution. They also need to consider the cost of selecting, implementing, and upgrading the groupware solution. In addition, they need to consider that groupware cannot adequately support corporate training, executive communication, or market research unless it is complemented by desktop videoconferencing.

3.5 CORPORATE NETWORKING SOLUTIONS

Currently, there are several distance networking solutions that can meet at least some of the telecom/datacom managers' requirements and that can support some of the IDL application solutions described in the previous section. These solutions include:

- Private DS1 lines;
- Switched/dialup services;
- SMDS;
- FRS;
- ATM/CRS.

In addition, several networking solutions are available to support the needs of telecommuters. These solutions include dialup analog services and ISDN. For each solution, a brief description is provided and its key strengths and limitations relative to IDL applications are highlighted.

3.5.1 Private Lines

A private line is a dedicated service that operates at DS1, DS3, and SONET speeds. This service provides a communication link between two locations through the establishment of a physical connection. Private lines are currently widely available in the United States and are offered by LECs, AAPs, and IXCs. As an IDL solution, the private DS1/DS3 solution provides a corporation with several benefits, including the following:

- *Support for IDL application solutions.* Private lines can support most IDL application solutions, including one-way video/two-way audio, two-way video/two-way audio, groupware, and the Internet.
- *Support for installed base.* Private networks became an integral part of corporate networks in the 1980s. The introduction or expansion of IDL solutions to a corporate networking environment may simply mean the allocation of a number of spare private-line channels to distance learning application solutions or the expansion of the installed base of private lines.
- *Security.* Private lines are perceived by telecom/datacom managers to be secure, since they are dedicated to individual customers. An increasing number of these lines are carried over fiber facilities.
- *Adequate bandwidth.* DS1 private lines provide corporations with adequate bandwidth. This bandwidth can be subdivided into multiple subchannels. Some of these channels can be dedicated to an IDL solution, while other subchannels can be used for other corporate applications. Each channel or group of channels can support various types of traffic, including voice, signaling, data, and video.
- *An established supply structure.* Private DS1 line services are established services. They are widely supported by LECs, IXCs, and AAPs. In addition, the increasing competition is driving down costs and making the service even more appealing.
- *Reliability.* DS1 private lines are reliable and proven. Information transport in dedicated bandwidth time slots guarantees throughput with no lost packets or processor bottlenecks.
- *Scalability.* DS1 private lines are scalable (e.g., DS3 lines), allowing for flexible network configuration, particularly for small networks.

Many organizations have now moved up to DS3 dedicated lines, providing 45 Mbps, or the equivalent of 28 DS1 lines. Typically the corporate telecom/datacom manager allocates some of this bandwidth to data applications (both legacy mainframe traffic and legacy LAN traffic), voice applications (PBXs), and video.

Private networking solutions are not, however, optimal in all situations. One of the drawbacks is that DS1 economics become less appealing to tele-com/datacom managers as the number of corporate locations requiring inter-connection increases. TDM-based private-line solutions are also bandwidth inefficient because they allocate bandwidth to devices that have nothing to send. Another shortcoming of DS1 private lines is that as private networking solutions they cannot support the IDL networking links of corporations with universities, high schools, suppliers, partners, customers, and telecommuters.

An example of a private line–based videoconferencing network is that of-fered by SBC [8]. This service connects corporate locations in Oklahoma with IDL providers, such as Oklahoma State University. Through the service, em-ployees of companies such as Conoco and Philips Petroleum can enroll in uni-versity graduate courses in petroleum, electrical, and mechanical engineering. This IDL solution provides corporations with a recruiting edge and with the ability to enhance the skills of their employees, while saving those employees the time associated with traveling to a university location, which could be as far away as 60 miles.

3.5.2 Switched/Dialup Services

Switched/dialup digital services are circuit switched–based services that sup-port speeds ranging from 56 Kbps to DS3 and support voice, video, and data traffic. With private lines, corporate participants need multiple dedicated lines to establish connections with multiple locations; with switched/dialup service, corporate participants need only one access line per required connection; with a network-resident bridge, only one line per location is needed (see Figure 3.3). To establish a connection with other locations, a corporate participant dials a number (call setup is based on the North American dialing plan). When the circuit switch receives the call request, a circuit is established for the duration of the call. This circuit is taken down as soon as the call is completed.

Switched/dialup services offer corporate participants several advantages over private lines:

- *Support for IDL application solutions.* Switched/dialup services can sup-port most IDL application solutions, particularly at the higher speeds, in-cluding one-way video/two-way audio, two-way video/two-way audio, and the Internet.
- *Cost reduction.* Switched/dialup services could be more economical than dedicated DS1 lines, particularly when there are many sites to be inter-connected, because with switched/dialup services each corporate site re-quires only one connection to reach all other destinations when a bridge is used (otherwise, multiple access lines are required). Switched/dialup

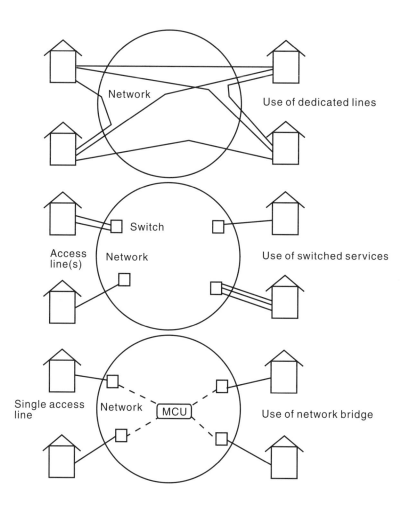

Fifgure 3.3 Dedicated versus switched connectivity.

services can also reduce corporate administrative costs if network connections of a corporation cannot be predicted in advance, or if these connections are constantly changing. Switched/dialup services also simplify network administration because they use the E.164 ISDN numbering plan. Switched/dialup services may also be more economical than private lines because they are usage sensitive, while private-line services are not. The economics depend on usage levels; at low usage, dialup lines are cheaper.

- *Extensive bandwidth capabilities.* Switched/dialup services provide the telecom/datacom manager with bandwidth capabilities that range from 56 Kbps to 45 Mbps. However, a set of dedicated access lines is required.

- *Flexibility.* Switched/dialup services are flexible because they enable telecom/datacom managers to include additional locations in their corporate networks without having to justify the expense of an end-to-end dedicated line.
- *Migration capabilities.* Switched/dialup services provide corporate participants with a migration path to more advanced switched services such as FRS and CRS.

While switched/dialup services can provide a corporation with several benefits, they have several limitations:

- They are not widely available yet, particularly at the higher speeds (this situation may change in the next few years, because RBOCs are increasingly deploying circuit switched services at DS1 speeds).
- As a circuit-switched technology, switched/dialup services have a relatively long call-setup time (3 seconds), although that should not be a problem for any IDL applications except data distribution.
- Switched/dialup services cannot support LAN interconnection applications, such as groupware, in the most effective manner, because bridges can be unstable when interconnected through switched/dial up services.

3.5.3 SMDS

As discussed in Chapter 1, SMDS is a connectionless WAN service that supports multiple-access classes. One of these access speeds can be supported by a DS1 access line, while the other speeds (4 Mbps, 10 Mbps, 16 Mbps, 25 Mbps, and 34 Mbps) can be obtained by using a portion of a DS3 line. SMDS is also available at 64 Kbps. SMDS offers low delay (5–10 ms for a DS3–DS3 link), high throughput (95% of packets are delivered in less than 20 ms), low error rates, high reliability, and availability (99.9%).

SMDS addressing scheme complies with ITU-T E.164 plan for ISDN/BISDN "telephone numbers," providing end users with transition to ATM, at least in terms of this feature. SMDS has several service capabilities, including multiple addresses per interface, and authentication of address in each packet. An SMDS customer can choose a list of addresses to communicate with; all other addresses are blocked. In addition to its addressing capabilities, SMDS enables corporations to establish *logical private networks*, which could be used by corporations to establish connectivity with a select number of corporate locations, suppliers, or customers. SMDS also has multicasting capabilities, which are similar to the broadcast/multicast capabilities of LANs. These capabilities are useful for address resolution, router updates, and resource discovery.

Corporations can deploy SMDS either as a new service or as a replacement of private lines. To deploy SMDS, a telecom/datacom manager needs to order the service from the local exchange carrier, install a DS1 or DS3 SMDS access line from the desired location to the service office of the local exchange carrier, purchase a router or an SMDS upgrade for an existing router, and install a DSU/CSU on the access line. SMDS is principally used for data.

SMDS provides corporations with several benefits, including the following:

- *Cost reduction.* SMDS could be more economical than dedicated DS1 lines, particularly when there are many sites to be interconnected, because with SMDS each corporate site requires only one access connection to reach all other destinations. SMDS can also reduce corporate administrative costs if network connections cannot be predicted in advance, or if those connections are constantly changing (of course, each potential remote location needs to have an access line). SMDS is also more economical for small organizations that have considerable data communications needs but cannot justify the expense of building a private network. SMDS could also be viewed as an economical service, because it provides a relatively graceful migration path to ATM technology. This comes about since SMDS and ATM have several technical similarities, including the ability to transfer data in fixed, 53-octet cells and the E.164 address plan.

- *Support for bursty data applications.* SMDS is well suited for corporations that want to implement data-oriented IDL solutions, such as groupware and the Internet.

- *Extensive bandwidth capabilities.* SMDS provides the telecom/datacom manager with a choice of access classes, enabling the manager to start at a lower bandwidth level and then migrate the network to a higher access class as the communications needs of the institution warrant the upgrade.

- *Flexibility.* SMDS is flexible because it enables telecom/datacom managers to include corporations in their corporate networks that do not justify the expense of a dedicated line.

- *Security.* SMDS has several appealing security features, enabling telecom/datacom managers to establish virtual private networks. Addresses can be screened so that only authorized destinations can receive data, and only authorized sources can send data. Through the group addressing capabilities, corporations can broadcast data to multiple locations at once.

While SMDS can provide a corporation with several benefits, it has several shortcomings: It supports only packet data traffic, and it cannot support voice and real-time video applications. Consequently, SMDS cannot support a number of IDL solutions, including one-way video/two-way audio and two-way video/two-way audio.

3.5.4   Frame Relay Solutions

Frame relay networks provide another data communications alternative for corporations. FRS is a (mostly) data-only, connection-oriented frame-transport service over assigned virtual connections. These virtual connections, which are supported through statistical multiplexing techniques, can be either permanent or switched. The FRS available on the market at press time was PVC based. A PVC connection is usually established when the service is provisioned at subscription time, eliminating the need for user-to-network signaling. FRS supports access speeds of 56 Kbps, $n \times 64$ Kbps, and 1.544 Mbps.

A frame relay network is based on ANSI standards, including T1.606, which specifies user-to-network interface requirements; ANSI T1.617 annex D, which specifies network management functions; and ANSI T1.618-1991 (LAP-F Core), which operates at the lower sublayer of the data link layer and is based on the core subset of T1.602 (LAP-D). The frame relay data transfer protocol defined in T1.618/LAP-F Core is intended to support multiple simultaneous end-user protocols within a single physical channel, since, above layer 2, this protocol is transparent. As a result, most existing protocols (e.g., TCP/IP) can ride over frame relay transparently to the end devices.

Frame relay can be implemented as either a private or a public solution. Frame relay can be implemented as a private networking solution by either (a) purchasing routers (or frame relay software additions to existing routers) and establishing point-to-point connections among those routers, or (b) using a customer-owned frame relay nodal processor (or a software upgrade to an existing packet switches or multiplexers), which is basically a frame relay switch.

Corporations can access a public frame relay network by (a) installing a software upgrade to a router or a bridge or purchasing a *frame relay access device* (FRAD) or direct firmware support on a host; (b) connecting an access line between the carrier and the corporations' CSU/DSU; (c) ordering FRS from a carrier offering the service; and (d) configuring PVCs for the sites to be linked to the network. A gamut of carriers currently support FRS, including RBOCs, the major IXCs, and several VAN providers.

FRS provides corporations with several benefits, including the following:

- *Economic benefits.* FRS offers a more economical networking solution than TDM-based DS1 lines because it is based on statistical multiplexing techniques. In the long term, FRS could also be viewed as an economic service, because it provides a migration path to ATM in the sense that it gets the organization on a packet-based paradigm.
- *Support for bursty data traffic.* FRS is well suited for specifically supporting data-based IDL application solutions, such as groupware and Internet access.

- *Availability*. FRS is widely available and is currently supported by IXCs, LECs, and RBOCs. FRS is also increasingly supported by the AAPs, such as Teleport Communications Group.

While FRS can provide a corporation with several benefits, it has a number of shortcomings, including:

- *Lack of support for isochronous traffic*. FRS supports only bursty packet data traffic. It cannot effectively support voice and real-time video applications. Consequently, FRS cannot support several IDL application solutions, including one-way video/two-way audio or two-way video/two-way audio.
- *Complex administration*. FRS requires careful administration for large PVC networks because the network administrator must define the connections for everyone on the network, including users who are added or changed. Network administration is becoming an increasingly important issue for large corporations, which are witnessing two conflicting but co-terminous trends: network consolidation used as a cost-cutting measure and network expansion as a result of mergers and acquisitions or a reach for complete distributed computing (e.g., client/server).
- *Bandwidth limitations*. PVC FRS offerings support speeds of up to DS1 (E1 speeds in Europe). As result, frame relay may not be able to support the bandwidth requirements of technical IDL receivers and providers, for example, quality video.
- *Transmission delays*. FRS, which transmits information in relatively small data units, has slightly higher end-to-end delay compared to private lines of the same bandwidth because of the need to store and process packet headers and trailers. FRS are also prone to queuing delays because multiple sources are competing for a given trunk circuit.
- *Packet loss*. Higher layer protocols have to recover lost data.

3.5.5 ATM Networking

ATM is a high-bandwidth low-delay switching and multiplexing communication technology that supports both LAN and WAN communications. To be exact, ATM refers to the network platform, while CRS refers to the actual service obtainable over an ATM platform. It is the general industry consensus that ATM is the service of choice for multimedia and other high-capacity interactive video-based applications [9]. For readers familiar with the operation of a protocol stack, it is simply a matter of realizing what functional partitioning has been instituted by the designers of ATM and what are the peer entities in the user's equipment and in the network. The cell relay protocols approximately correspond to the functionality of the medium-access control layer of a traditional

LAN but with the following differences: random access is not utilized, channel sharing is done differently, and the underlying media may be different.

Connections in an ATM network support both circuit mode and packet mode services of a single media and/or mixed-media and multimedia. As covered in Chapter 2, ATM carries three major types of traffic: CBR, VBR, and ABR. For example, traditional video transmission (whether compressed or not) generates CBR traffic, while data applications (say, router traffic for a traditional LAN) generate VBR traffic.

Two remotely located corporate multimedia or video-based devices required to communicate over an ATM network can establish one or more bidirectional virtual (i.e., not hard-wired or dedicated) connections between them. That connection is identified by each user by an appropriate identifier. Once such a basic connection is established, user devices can utilize the virtual, connection-oriented channel for specific communication tasks. Each active channel has an associated bandwidth negotiated with the network at connection setup time. The transfer capacity at the *user network interface* (UNI) 155.52 Mbps; other interfaces are also being contemplated in the United States at the DS1 (1.544 Mbps) and DS3 (44.736 Mbps) rates.

The ATM architecture utilizes a logical protocol model to describe the functionality it supports. The ATM logical model is composed of a user plane, a control plane, and a management plane. The *user plane*, with its layered structure, supports user information transfer. The *control plane* also has a layered architecture and supports the call control and connection functions; it deals with the signaling necessary to set up, supervise, and release connections. The *management plane* provides network supervision functions. It provides two types of functions: layer management and plane management. Plane management performs management functions related to a system as a whole and provides coordination among all planes. Layer management performs management functions relating to resources and parameters residing in its protocol entities (see Figure 3.4).

In the user plane, the access protocol in the user equipment consists of a physical layer at the lowest level and of an ATM layer over it that provide information transfer for all services. Above the ATM layer, the *ATM adaptation layer* (AAL) provides service-dependent functions to the layer above the AAL. TCP/IP may continue to be used by users' PCs and hosts (note that AALs usually only go as high as the data link layer). There are currently three AAL protocols: AAL 1, AAL 3/4, and AAL 5. The service data units reaching the AAL consist of user information coming down the protocol stack, for example, from a TCP/IP stack or from a video codec; the information is segmented or cellularized by AAL into the 53-octet cells, so that they can be efficiently shipped through the network. For video and/or multimedia applications, AAL 1 (supporting CBR) can be employed by the user equipment; more recently, however, many have also advocated the use of AAL 5 for video.

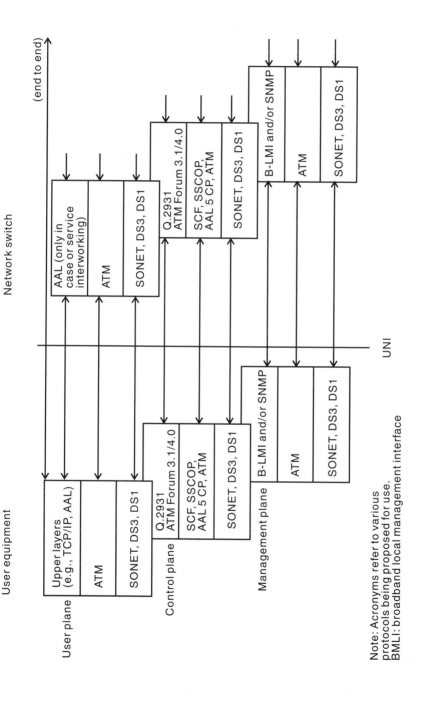

Figure 3.4 ATM protocol planes.

The AAL enhances the services provided by the ATM layer to support the functions required by the next-higher layer. The AAL-specific information is nested in the information field of the ATM cell. To minimize the number of AAL protocols, a service classification was defined in the late 1980s based on the following three parameters: timing relation between source and destination (required or not required); bit rate (constant or variable); and connection mode (connection oriented or connectionless). Five classes of applications were then defined, as follows:

- *Class A:* Timing required, bit rate constant, connection oriented;
- *Class B:* Timing required, bit rate variable, connection oriented;
- *Class C:* Timing not required, bit rate variable, connection oriented;
- *Class D:* Timing not required, bit rate variable, connectionless;
- *Class X:* There are no restrictions (bit rate variable, connection oriented, or connectionless).

Class A is similar to a circuit emulation (dedicated line) service. AAL 1 is used in conjunction with Class A service. A network supporting pure CRS, namely, the orderly, reliable, expeditious, high-throughput movement of user cells from an origination interface to a destination interface, supports Class X applications.

CRS can be used in a variety of ways to support multimedia and video-based applications. In supporting multimedia applications, several issues depend on whether one uses multiple "single-media" connections (performing multiplexing at the ATM layer) or a single multimedia connection (performing media multiplexing above the ATM layer). Some may contemplate using a single VBR cell relay connection for multimedia; here all media multiplexing is done by the application, using proprietary protocols at the AAL layer running over a network providing pure CRS. Another approach is to do multiplexing at the ATM layer, enabling the carrier to provide desirable value-added features. Figure 3.5 depicts a typical use of ATM in a video environment, for example, to support IDL, from a protocol perspective point of view.

More recently, the CBR/VBR/ABR/UBR view discussed in Chapter 2 has been advocated or adopted to describe the various services made available by ATM.

Multipoint connectivity is generally an important requirement for IDL. This follows from the fact that the instructor needs to be seen and heard by a (large) number of remote sites. Satellite transmission supports multipoint very well. However, terrestrial networks do not (yet) support that requirement in an effective manner. This fact has given impetus to the development of ATM/cell relay service: Point-to-multipoint and multipoint-to-multipoint connectivity will be supported as a key service feature.

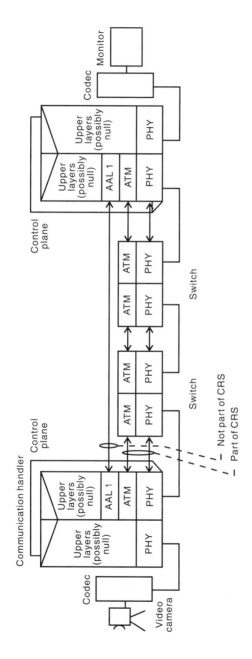

Figure 3.5 Using ATM to support video-based instruction.

ATM/CRS can provide corporations with several benefits, including the following:

- *Support for data, voice, video, and multimedia traffic.* CRS provides corporations with the versatility and flexibility of fixed-length cells, making the service well-suited for all IDL solutions because of the low latency incurred in switching.
- *Economic benefits.* CRS offers higher transmission speeds than SMDS and FRS. It is also more economical than TDM-based lines because a single access line is required to establish communications links between corporate locations; with private lines, multiple access lines are required. Generally, carriers are pricing ATM/CRS aggressively.

While CRS can provide a corporation with several benefits, CRS based on PVCs has one shortcoming: As a PVC service, CRS requires careful administration for large networks because the network administrator must define the connections for everyone on the network. SVC services will be added in the 1996–97 time frame.

3.5.6 Dialup Solutions

Analog dialup solutions can support the needs of corporate telecommuters for remote university education. For example, low speed is employed for NYU's Virtual College. Using their home personal computers and modems, corporate employees, who are university students, can receive instruction, ask questions, conduct analyses, resolve problems, and prepare professional studies—all largely at their own pace and convenience. The Virtual College is an electronic learning environment for the efficient production and delivery of a wide range of high-quality business and technical courses. NYU's Virtual College and its on-line educational network are designed to meet these expanding training needs in an efficient and effective fashion. The program employs Lotus Notes software. Lotus Notes is a group communications program (groupware) giving people who work together an electronic environment within which to create, access, and share information, using networked personal computers. Lotus Notes support such business applications as computer conferencing, information distribution, status reporting, project management, and electronic mail.

To access a course offered by NYU's Virtual College, students must have an IBM-compatible PC with Microsoft Windows, a minimum of 2 MB of memory, 8 MB of available hard disk space, 1.44-MB 3.5-inch diskette drive, VGA monitor, and a 2,400- or 9,600-bps Hayes or Hayes-compatible modem. Optionally, users can access the telecourse from a work location using a LAN-based PC that either has direct-dialout capability or utilizes a communication server (modem pool). However, there is a trend up to ISDN. This topic is examined further in Chapter 10.

3.5.7 ISDN

ISDN is a networking solution that can support the IDL needs of corporate employees through digital switched channels operating at speeds that fill the gap between traditional analog dialup and dedicated or switched DS1 lines. An example of an ISDN-based solution is shown in Figure 3.6, which depicts a telecommuting/distance learning application. This solution provides the telecommuter with the ability to participate in a LAN-based screen-sharing conference that is taking place at a corporate location and to conduct a voice conference with the LAN participants as well. The voice conference is established using an existing non-ISDN conference bridge. The LAN-connected participants then establish screen-sharing sessions with the LAN conference bridge. The LAN conference bridge supports multipoint conferencing functions. As a result, the software on each participant's PC only needs to handle a point-to-point conference with the bridge. The telecommuter establishes a circuit-data connection to the communication server that makes the computer appear to be on the LAN. The telecommuter then runs the same software as to set up a conference with the LAN conference bridge.

As Figure 3.6 shows, to establish an ISDN link with the corporate office, a user needs an analog phone, which enables the telecommuter, acting as a conference controller, to establish a voice conference using a non-ISDN conference bridge. The user also needs a PC or workstation equipped with an ISDN adapter, applications software compatible with software on the corporate site. This PC also has several B-channel capabilities, including simultaneous voice and circuit-data calling, support for two directory numbers and service profile ID, and support for rate adaptation for rates lower than 64 Kbps.

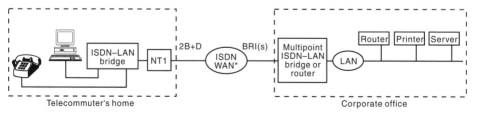

* WAN = Wide area network

Figure 3.6 Telecommuter/corporate office link: the ISDN solution.

On the corporate side, a communications server is required. This server must have application software compatible with communications software and with far end. This server must also have multiple BRI capabilities, the same B-channel capabilities as the PCs. A LAN conference bridge and PCs are also needed at the corporate site. In addition, the LAN conference bridge must have must multipoint conferencing functions to support four or more users.

3.6 SUMMARY

IDL can play a vital role in addressing the following key challenges of corporations: corporate training, intracompany collaborative learning and research, remote university education, intercompany collaborative learning and research, executive communication, and market research. These applications can be addressed by a diverse range of application solutions, as shown in Table 3.7. A number of networking solutions can support each application solution and are shown in Table 3.8.

Table 3.7
Corporate Distance Learning Application Solutions

	Corporate Application Solutions				
Application Solutions	Corporate Training	Executive Communication	Intracompany Collaboration	Intercompany Collaboration	University Education
Business TV	×	×			×
Two-way video conferencing	×	×	×	×	×
Groupware			×	×	×
Internet			×		×

Table 3.8
Corporate Distance Learning Application/Networking Solutions Mix

	Corporate Networking Solutions					
Application Solutions	*Private Lines*	*Circuit Switched*	*ISDN*	*FRS*	*ATM*	*SMDS*
Business TV	×	×	×		×	
Video conferencing	×	×	×		×	
Groupware	×			×	×	×
Internet	×	×	×	×	×	×

References

[1]   Eldib, O., and D. Minoli, *Telecommuting*, Norwood, MA: Artech House, 1995, p. 25.
[2]   "Ford Uses Distance Learning, Cray Supercomputers To Keep Engineers Current," PR News Wire., Feb. 16, 1994, p. 216.
[3]   Tucker, Tracey, "Video Update; How Things Are Shaping Up and What's on the Horizon," *Teleconnect*, May 1993, p. 98.
[4]   NATA, "1993/1994 Telecommunications Market Review & Forecast," Section VII-4, 1993, p. 178.
[5]   *Videoconferencing*, Faulkner Technical Reports, Inc., 780.0000.100, August 1993, p. 4.
[6]   Minoli, D., "WilTel/Vyvx Video Services," *Datapro Report CNS*, June 1995.
[7]   "Corporate Classroom Provides Career Opportunities," Managing Office Technology, Cleveland, Ohio, June 1993.
[8]   Minoli, D., "ATM Makes Its Entrance," *WAN Connections*, suppl. *Network Computing Magazine*, Aug. 1993, pp. 22 ff.

Bibliography

"Big Business TV Network Provides Boost for Ford," *Communications Week*, March 1994, p. 51.

Lambert, P., "Distance Training Saves Minnesotans Time and Money," *Multichannel News*, Oct. 4, 1993.

Distance Learning Needs of Universities

4

Universities play two roles relative to IDL; they are both users and providers of IDL services. This chapter explores both roles as well as the educational challenges facing universities. It then identifies the IDL applications that can meet those challenges, and the university needs associated with each application. The chapter then identifies and describes key solutions at the application level and the networking level. The chapter concludes by positioning the various solutions against each other relative to the needs of universities.

4.1 CHALLENGES FACING UNIVERSITIES

Universities are faced with a number of major challenges. First, if they follow the traditional approach in delivering education, they must continue to incur the cost of building new physical infrastructures, such as campuses, dormitories, and libraries. They also have the onus of maintaining existing campuses. Another key challenge facing universities is the increasing cost of education. University tuition is constantly on the rise, outpacing inflation. The average tuition reached $11,704 at four-year private colleges in 1994 and $2,686 at four-year public colleges, both up 6% over 1993. When room, board, books, supplies, and transportation are included, the total cost to resident students at four-year private colleges amounts to $18,784 and $8,990 at public ones (students usually pay about half the college bill [1]).

A third problem facing most universities in the United States is the declining interest of some segments of the graduating high school population in obtaining a college education. This is forcing universities to compete in order to attract graduating high school students. One of the tools that universities are using to attract students is to offer lower tuition. For example, LeHigh University has cut tuition 22% for its master's program. M.B.A. graduates are also entitled to get two-thirds on the regular tuition for any course they take after graduation. The University of Rochester announced that it would give a $5,000 grant to all in-state students who enrolled as freshmen in 1995. The University of Detroit Mercy offers out-of-state students a grant program of up to $1,950 to match the amount they receive from Michigan students.

A fourth problem facing universities is the increasing budgetary constraints facing the government at the federal and state levels. This has forced state governments to proposed major cuts in their support for state universities.

While universities are attempting to attract students and to deal with tighter budgets, their success has been limited by the need to repair and replace aging buildings. Some universities are also spending more on their R&D efforts to maintain their roles as R&D service providers to industry and government.

4.2 DISTANCE LEARNING LINKS AND APPLICATIONS

The problems highlighted here represent substantial challenges facing universities and colleges. IDL is one the technological tools that can contribute to the solution of these educational problems. IDL can provide universities with the following benefits:

- IDL links enable universities to establish closer ties with their corporate and government clients by enabling them to rapidly share information among members of the R&D community, speeding up the process of R&D and the development of new products and services.
- IDL can also resolve some of the financial problems facing universities by providing them with economies of scale. Through IDL, universities can create a greater number of remote classes. This provides education providers with the opportunity to generate more revenues per teacher or to reduce the tuition per student.
- A university can differentiate itself from other universities by being at the leading edge of IDL technology.

Universities, however, cannot hope to maximize the benefits of automation without attacking those problems through specific IDL applications, identifying the communications applications that are driving the problems, and implementing innovative solutions that can address the problems.

Figure 4.1 depicts a distance learning communications model that is applicable to universities. The scope of the model is limited to intrauniversity and interuniversity communications links, excluding university links with corporations and K–12 schools (those links are examined as part of the IDL communications model of corporations and K–12 schools discussed in Chapters 2 and 3). As Figure 4.1 shows, various university buildings need to establish intrauniver-

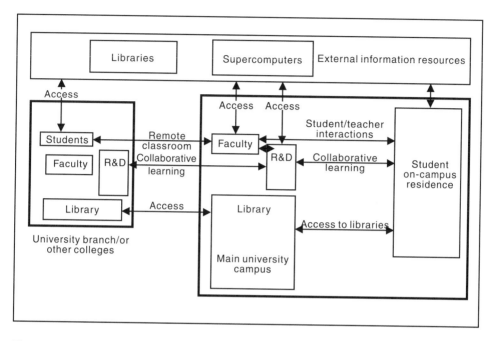

Figure 4.1 Major university distance learning applications.

sity and interuniversity communications links with university branches, other colleges, dormitories, and external sources of information.

The following IDL applications drive universities' needs for the establishment of these links:

- Collaborative learning and research;
- Access to supercomputers;
- Access to libraries;
- Remote education;
- Access to databases.

4.2.1 Collaborative Learning

The university community relies on the sharing of knowledge as the key tool to discover new technologies and to educate its students. Collaboration is important to university students so they can share and analyze scientific data, have extended discussions with other peers, and work together to produce papers and journals.

Traditionally the sharing of information relied on text-based e-mail and on face-to-face meetings. Increasingly, university students and researchers are seeking (and sometimes getting) computer graphics and animation-generation systems with text. For example, oceanographers [2] need to transmit images in digital form to allow remote researchers to work with a file. That, however, requires the availability of broadband networks that support data, text, and image communication.

4.2.2   Access to Databases

University students, instructors, and researchers need to have access to a large number of databases. The nature and locations of the databases are shaped by the type of research that an individual is performing. Examples of those databases include the GENBANK and the Cambridge Crystallographic Database serving the needs of the medical and chemistry research communities, the human genome sequence, the global seismic databank, and economic and social databases.

These databases have two common characteristics: (a) They are accepted standards in their respective research communities, and (b) they are large. The large size of the databases and the impracticality of duplicating them in every research lab across the nation necessitate the availability of broadband networks to researchers to maximize the speed of scientific research and development.

4.2.3   Access to Supercomputers

Access to supercomputers is another important IDL application, for example, for visualization purposes. University needs for access to supercomputers are shaped by the needs of the various groups of researchers operating in an university environment. Following are a few examples of the applications that drive the needs of university researchers for access to supercomputers:

- Chemists need remote supercomputer access for the generation of electron-density maps by chemical modeling.
- Researchers in mechanical engineering need supercomputer access and communications images of three-dimension images in major areas such as aircraft simulations, vehicle simulations, turbulent fluid simulations, simulations of noise generation by vehicles and aircraft, and simulations of the particles mechanics of soil.
- Structural engineers need to access supercomputers to design major structures.
- Researchers in chemical engineering need supercomputer access for simulations of thermodynamic properties of liquids and gases.

For collaborative learning to be effective, educational experts believe that schools have to change in the following ways:

- Effective collaboration becomes more useful as the number of teachers and students participating in a specific collaborative effort decrease.
- Effective collaboration among students and teachers belonging to the same team requires more effective scheduling for groupwork. This effective collaboration means that interactions among members of a collaborative team need to be expanded beyond school hours to encompass interactions.

Broadband networks can address those issues.

4.2.4 Access to Libraries

Access to libraries is an important data- or image-oriented IDL application of university students. In preparation for their assignments, students need to access digital libraries as a substitute for physical access to libraries. This will provide them with access to a wider range of information resources that may be available in their university library. They need connectivity to these libraries to access electronic books, periodicals, and other publications.

4.2.5 Collaborative Learning and Research

Collaboration is critical to research and learning in the university environment. Collaboration is the driver behind many of the academic discoveries. As shown in Figure 4.1, collaborative learning encompasses the establishment of communications links among students, teachers, and researchers. Collaboration is not limited to the formal hours of schools, nor is it limited to a university campus. Instead, collaboration can occur 24 hours a day. Its physical boundaries extend beyond the university campus to encompass the world.

Students can benefit from getting in touch with their professors after class hours for several reasons: (a) to inquire about a specific project or report assignment, (b) to deliver the work to their instructors electronically, and (c) to receive comments back from their professors.

4.2.6 Remote Learning

Traditionally, universities needed to create remote classes to support the educational needs of their smaller branches as well as those of smaller universities. Remote classes provide universities with the following benefits: (a) they increase the number of teachers per student, and (b) they reduce the costs associated with building new classes and expanding campuses. While these were traditional benefits of providing IDL services, a more urgent benefit of IDL to universities is now *competitive advantage*. Universities are forced to increasingly operate as businesses; this requires them to balance more carefully their social goals with their key business goal of remaining viable and profitable.

To create a successful remote education solution, universities need to consider the following needs of students:

- *Interactivity*. This is a key element of any successful IDL program. The more the IDL solution selected by the university involves interactivity, the more it will be successful in getting students involved, enabling them to ask questions and to be stimulated while taking the course.
- *Instructional feedback*. This extends beyond the provision of grades by instructors to students on specific assignments, and it encompasses providing answers to questions after university hours.
- *Elimination of time constraints*. Through IDL solutions, students and instructors can, to some extent, free themselves from the constraints of time. This means that an effective IDL solution should enable students to maximize the time of their interactions beyond classroom hours.
- *Motivation*. An effective IDL solution should enable students to have fun as they interact with other students and teachers.

4.3 THE DISTANCE LEARNING NEEDS OF UNIVERSITY IS ADMINISTRATORS

University needs for IDL are shaped not only by the needs of students, instructors, and researchers but also by the needs of the telecom/datacom managers, who are responsible for the evaluation, selection, design, implementation, and operation of IDL solutions. In selecting the IDL applications, telecom/datacom managers of universities need to take into account a number of factors, some of which are specific to the university telecom/datacom environment.

- *The university information infrastructure*. The telecom/datacom infrastructure of a university is shaped by several factors: (a) whether the university or college is public or private, (b) the founder of the university, (c) the size of the university, (d) the specialization of the university, and, (e) the physical location of the university (whether it is an urban or suburban setting). The telecom/datacom environments of large universities resemble those of large industrial parks. The university telecom/datacom infrastructure may include some or all of the following elements:
  - *A voice networking environment*, supported by a PBX or centrex and extended over a metropolitan area through either private lines or a public switched network system.

- *Legacy systems*, which include older minicomputers, administrative mainframes, and even Ethernets and token rings.
- *A client/server environment.* Clients and servers, which may be spread around the campus and interconnected across LANs, such as Ethernet, token ring, and FDDI. Clients and servers may also be in different locations and interconnected.
- *Satellite-based networks,* used by universities both to originate and to receive IDL programming.

While universities may have similar infrastructures as those built by corporations, they differ in one major way: Universities are innovators and are willing to explore new technologies, even if those technologies are not well proven. In contrast, corporations would consider only a proven technology before committing their resources to the adoption of a technology.

- *Geographic coverage.* Telecom/datacom managers of universities seek application solutions that support the geographic range of universities. The geographic scope of universities may span a city or a state.
- *Ease of use and administration.* Telecom/datacom managers value applications solutions that are easy to administer and to use by IDL receivers and providers.
- *Intraenterprise versus interenterprise communications.* The appropriateness of the application solution to a university may depend on the extent of the needs of telecom/datacom managers for interuniversity versus intrauniversity communications.
- *Interoperability and compatibility.* Telecom/datacom managers are seeking solutions that are not proprietary but based on open standards. This would enable them to not have to rely on a single vendor.
- *Support for installed base.* Telecom/datacom managers value solutions that can build on their installed base of information systems.
- *Access methods and speeds.* Telecom/datacom managers have needs for different access methods and speeds in support of applications solutions. Those requirements are shaped by the installed base as well as by cost factors. The range of access methods and speeds required by telecom/datacom managers in support of application solutions may differ from one school to another. Access speeds may range from 1.2 Kbps to 155 Mbps.

4.4 DISTANCE LEARNING APPLICATION SOLUTIONS FOR UNIVERSITIES

A number of application solutions are available to universities in addressing their IDL needs: the Internet, groupware, one-way video/videoconferencing, and two-way video.

4.4.1 The Internet

Universities have been one of the initial drivers behind the growth of the Internet and have been using the Internet for over a decade. Typical uses have included:

- *Access to libraries and databases.* Through the Internet, university students can access hundreds of libraries around the world, as well as library catalog and full text-service delivery services. The Internet also enables university students and researchers to access a wide range of government agency information, such as space shuttle updates. In addition, the Internet provides participants with a wide range of public domain software and freeware documents, databases, images, and other files that they can rely on in their technical development efforts.
- *Collaborative learning.* Students can rely on a number of Internet services in collaborating with each other. These services include e-mail, logon services, file transfer, host-to-host communications, and directory services. These applications enable university students to establish a dialog with other researchers and scientists around the globe. It also enables them to transfer large files. In addition, through *electronic whiteboards* university students can share in real-time notes about specific technical topics. This increasing level of interaction among university students and researchers is speeding up the process of scientific research and greater contributions from universities to the discovery of new technologies.
- *Remote learning.* One of the growing capabilities of the Internet is videoconferencing. As mentioned in Chapter 2, the Internet has demonstrated the ability to link multiple sites though entry-level two-way videoconferencing links.
- *Access to supercomputers.* The Internet also enables university students and researchers to access major supercomputer centers throughout the country. The recent upgrade of the Internet backbone to a SONET/ATM platform will provide researchers with additional bandwidth facilitating the transfer of files to supercomputers.

The Internet will continue to be a key networking and communications solution for universities due to its low cost, support for a wide range of applications, and ability to provide university students and researchers with access to a range of information resources. In addition, the Internet solution provides students, researchers, and teachers with freedom from the constraints of time and space, interactivity, and instructional feedback.

The Internet has one major limitation, however, that universities need to take into account in selecting the applications supported by the Internet: security. Violations result from unauthorized password use, particularly given the emergence of programs collecting names and passwords on the network. Because of these security concerns, universities have to be cautious about using the Internet to serve the R&D needs of specific clients that do not wish to share their secrets with others.

4.4.2 Groupware Solutions

Groupware represents an emerging alternative solution to the Internet in meeting the needs of universities for intrauniversity IDL applications. The following key intrauniversity applications are supported by groupware (see also Chapter 10):

- *Collaborative learning and research.* Through groupware messaging applications, university students can submit their homework to teachers via e-mail, leave messages, or inquire about specific problems they are encountering in solving a specific homework assignment. Groupware enables students to collaborate in real time on projects. For example, they can view, modify, and approve documents by opening windows that contain a "whiteboard" on their individual PCs and making individual contributions as though they were using different colored "markers."
- *Access to databases.* Through groupware, students can access public databases, private databases, fax, paging, and voice mail.
- *Remote classroom.* Groupware can complement desktop videoconferencing in conducting small classrooms. Through a small window on desktop workstations, university students are able to establish videoconferencing sessions with remote teachers.
- *Student/teacher and student/student interactions after hours.* Through groupware, students residing in dormitories can exchange messages with teachers and other students, extending the collaborative research and learning efforts beyond school hours.

Groupware can also increasingly support interuniversity applications. To establish interuniversity distance links, universities need to rely on a public network service providers, such as AT&T, which offers a Lotus Notes–based public groupware solution.

Groupware offers universities several benefits, including the following:

- *Low cost.* In implementing groupware, telecom/datacom managers incur a nonrecurring cost. This cost is on either a user or a server basis.
- *Multiple applications.* Groupware supports multiple IDL applications, as described above.

- *Ease of use.* Multifunctional groupware solutions are based on GUIs, which makes them easy to use. Consequently, groupware solutions are appropriate for all groups of IDL receivers and providers.
- *Secure.* As a private solution, groupware provides a secure solution for intracompany communications.
- *Training and consulting.* The leading providers of multifunctional groupware provide their buyers with training or consulting.

While groupware offers universities a number of benefits, telecom/datacom managers should take into account the following considerations: (a) the cost of selecting, implementing, and upgrading the groupware solutions, and (b) groupware solutions cannot effectively support access to supercomputers or to databases.

The benefits of groupware outweigh its limitations. Consequently, the adoption rate of groupware solutions by universities will likely accelerate.

4.4.3   The One-Way Video/Two-Way Audio Solutions

One-way video/two-way audio is another IDL application solution that universities can (and do) rely on. Satellite-based or terrestrial-based business-TV solutions, such as those described in Chapter 3, support the establishment of remote classrooms, enabling university students in remote locations to watch a teacher though a one-way video link and to ask questions through a two-way audio link.

An example of a public-based business-TV solution for universities is the Mind Extension University, which was founded in 1987. The network combines the two technologies of satellite and cable TV to deliver over 150 credit courses per year at undergraduate level and 45 at the master's level.

The one-way video/two-way audio solution provides university students with the ability to ask questions and interact with other students in other locations. The one-way video/two-way audio solution is beneficial not only to university students but also to telecom/datacom administrators of universities for the following reasons:

- *A large number of satellite-based providers are currently available.*
- *A high-quality picture.* Satellite and fiber systems deliver high quality pictures.
- *Affordability.* The cost of one-way transmission is much less than the cost of hiring a teacher.
- *Support for interuniversity and intrauniversity networking solutions.*
- *Geographic coverage.* The satellite-based one-way video solutions has a regional or national geographic coverage. Similar reach can be obtained with fiber but this solution is more expensive.

All these factors make one-way video/two-way audio an appropriate IDL solution to university administrators. This solution, however, has its limitations: The major problem with one-way video conferencing systems to students is the accessibility of teachers. Consequently, the relationship that develops between the student and the teacher through face-to-face interactions cannot be duplicated through the one-way video solution. A third problem is scheduling complexities, because in many cases, the daily transmission hours are often at odds with a school's own class schedules.

The one-way video/two-way audio solution cannot be considered a total solution because it cannot support several IDL applications, including collaborative learning, access to libraries, and access to supercomputers.

4.4.4 Two-Way Videoconferencing Solutions

Two-way videoconferencing solutions provide a better alternative to the one-way video solution, particularly if it employs full-motion video. An example of a growing two-way videoconferencing solution is the *Collaboration for Interactive Distance Visual Learning* (CIDVL) [4]. This is a multi-institutional effort, which was formally launched in 1993. CIDVL consists of 11 universities and corporations seeking to create a "virtual university" environment, providing courses and lectures to one another in the field of engineering. Initial members of CIDVL were Boston University, MIT, Penn State, Rensselaer Polytechnic Institute, 3M Corp., AT&T, United Technologies, and PictureTel Corp. Other universities and corporations are entitled to join CIDVL as members by paying $10,000 per year for a three-year membership term. New members are also expected to offer their own intellectual contributions.

Students enroll in a CIDVL course through either their individual university or corporation. University students pay per-credit fees to the remote institution offering the program. Credit-transfer policies are arranged between the native university at which the student is enrolled and the remote university offering the course.

The CIDVL network is based on the PictureTel System 4000 family of videoconferencing products. Each participating institution must have access to the public switched 56-Kbps network or ISDN service. The PictureTel system transmits two-way video/audio at a compression rate of 1,000:1 (170 Mbps to 128 Kbps). Students see compressed images at 15 frames per second. The installation cost per site is approximately $40,000. CIDVL also makes videotapes of all the lectures available to students who want to review the lectures at a later time. The quality of the picture, and hence of the lecture, leaves much to be desired. In particular, the round-trip encoding delay frustrates interactivity.

The two-way videoconferencing solution described here offers a potential improvement over the one-way video/two-way audio solution: interactivity. With this solution, all the participants can see each other, enhancing the collaborative learning process. However, effectiveness depends on picture quality. In turn, the quality of the two-way videoconferencing picture, its affordability, and its geographic presence depend on the coding method and the networking solution provided. In the example mentioned here, the picture may not be of as high a quality as if the transmission were based on other solutions, such as FRS. The networking solutions will be explained in more detail in the next section.

Despite the benefits of this solution, it cannot on its own support all the IDL needs of universities. Data-oriented solutions, such as groupware and the Internet can complement the two-way video/audio solution in addressing the IDL needs of universities.

4.5 DISTANCE LEARNING NETWORKING SOLUTIONS

A number of networking solutions are available to support the application solutions described in the preceding sections. Key solutions include ATM-based solutions; fast-packet services, such as FRS and SMDS; and ISDN *primary rate interface* (PRI).

4.5.1 ATM Solutions

ATM is an emerging distance learning solution that can complement or substitute for other networking solutions, such as FRS, SMDS, and NMLIS. ATM can support *all* the IDL application solutions described in this chapter.

ATM can provide universities with high transmission speeds and the ability to integrate data, voice, video, and multimedia communications. One university that has implemented an ATM-based networking solution is Duke Medical Center in North Carolina. Administrators selected ATM as both a LAN and a WAN solution to replace an Ethernet-based LAN infrastructure (see Figure 4.2).

Before it implemented an ATM solution, the radiology department used Ethernet LANs as a shared-media solution, linking two separate hospital facilities, Duke North and Duke South. A bridge linked the radiology department's LAN to the public WAN, which supported other departments in Duke Medical Center. Another bridge provided a link to the LAN used to transfer downsampled (low-resolution) x-ray images. The Ethernet network had the following problems:

- To avoid broadcast storms, the amount of traffic allowed on the networks was limited. Consequently, not all radiology departmental LANs could be directly linked to the Ethernet network.

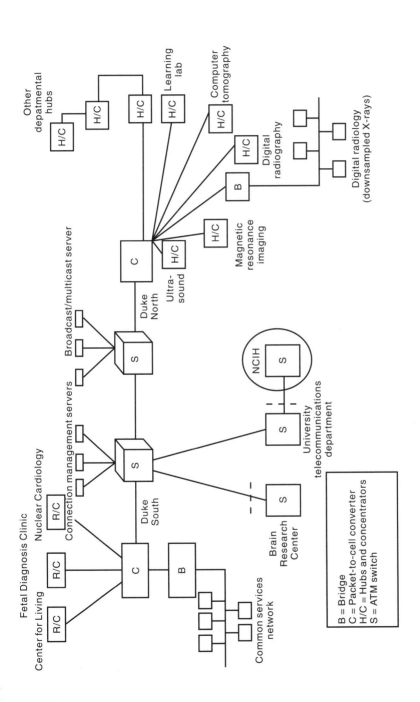

Figure 4.2 Duke University Medical Center ATM network.

- The radiology department could not download standard x-ray images, which comprise about 10 MB of data. Doctors requested 10 to 100 pictures at a time. Fulfilling a single doctor's request would have taken several minutes by relying on the Ethernet network. To accommodate as many doctors as possible, the radiology department had to use downsampled (low-resolution) x-rays, which could be used only for consultative purposes.

To overcome those problems, Duke Medical Center implemented a campus-wide ATM backbone connecting all departmental LANs in the radiology department. The backbone was also linked to an ATM switch, which in turn was linked to the NCIH. Through this network, doctors do not have to worry about bandwidth. Consequently, they are able to access x-ray images from their homes, downloading x-ray images from Duke Medical Center in a matter of seconds.

4.5.2 SMDS

SMDS could emerge as a distance learning solution that can complement or can substitute other networking solutions, such as FRS, NMLIS, and ATM CRS; however, its future may be limited due to the relatively rapid introduction of CRS, which supports data and video, and WANs and LANs. SMDS can support the data-oriented IDL application solutions described above, including access to the Internet and groupware communications.

In addition, SMDS provides universities with several benefits, including cost reduction, support for high-bandwidth bursty data applications, flexibility, and security. Each of these benefits is described next.

- *Cost reduction.* Initially SMDS could be more economical than ATM PVC solutions, particularly when there are many sites to be interconnected. SMDS can also reduce university administrative cost if network connections of a university cannot be predicted in advance, or if those connections are constantly changing. SMDS also provides a graceful migration path to ATM, since SMDS and ATM have several technical similarities.
- *Extensive bandwidth capabilities.* SMDS provides telecom/datacom managers with a choice of various access classes, enabling them to start at a lower bandwidth level and than migrate the network to a higher access class as the communications needs warrant the upgrade.
- *Flexibility.* SMDS is flexible because it enables telecom/datacom managers to include corporations in their networks that do not justify the expense of a dedicated line.

While SMDS can provide a university with several benefits, it has several shortcomings. Specifically, it is defined to support packet data traffic. Consequently, SMDS cannot support a number of IDL solutions, including one-way video/two-way audio and two-way video/two-way audio solutions.

A university that has implemented SMDS is George Mason University. The problem facing that university was how to: (a) upgrade their campus LANs from traditional Ethernet speeds to higher speeds and (b) connect Ethernet LANs of multiple campuses into a high-performance WAN. In solving that problem, the telecom/datacom management of George Mason University had the following requirements:

- Easy migration from a lower-speed Ethernet (10 Mbps) to higher speeds (at 100 Mbps);
- Maximization of bandwidth at the WAN level, considering that they were downloading executable portions of code, instead of just information;
- Ability to migrate their WAN to a higher bandwidth without having to redesign the WAN;
- Any-to-any connectivity;
- Minimum cost of upgrade.

To solve their problems, George Mason University selected switched Ethernet LANs at the LAN level and SMDS at the WAN level (see Figure 4.3). In selecting switched Ethernet LAN, telecom/datacom managers compared switched Ethernet to FDDI. The telecom/datacom management found that switched Ethernet had two advantages: (a) It was less expensive than FDDI, and (b) selecting FDDI as a campus backbone LAN would require translational bridging between an FDDI backbone and Ethernet segments, inevitably impairing performance.

In selecting SMDS, telecom/datacom management compared SMDS with FRS. SMDS had several advantages over FRS in the case of George Mason University: (a) Maximum bandwidth of FRS is DS1, while the selected SMDS service had a 10-Mbps speed; (b) telecom/datacom managers wanted any-to-any connectivity, which FRS could not deliver (a telecom/datacom manager would have had to call the carrier each time a new connection was established); and (c) SMDS provides an easier migration path to higher speeds. With SMDS, the university can migrate to speeds higher than 10 Mbps simply by changing the adapters on the routers. Universities making selections after 1995 may find that the increased availability of ATM simplifies their choice.

4.5.3 Frame Relay Solutions

Frame relay networks provide a data communications alternative for universities. FRS is emerging as a distance learning networking solution that can com-

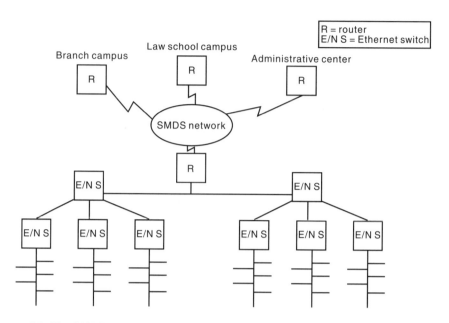

Figure 4.3 The SMDS network of George Mason University.

plement or substitute for other networking solutions, such as SMDS, NMLIS, and ATM CRS. As already noted, FRS is a data-only, connection-oriented frame-transport service that uses virtual connections. Statistical multiplexing techniques are used to support PVCs and, in the future, SVCs. FRS on the market tends to be PVC based. A PVC connection is usually established when the service is provisioned at subscription time, eliminating the need for user-to-network signaling. FRS supports access speeds of 56 Kbps, $n \times 64$ Kbps, and 1.544 Mbps.

FRS can support the data-oriented IDL application solutions described above, including access to the Internet and groupware communications.

Frame relay provides universities with several benefits, including the following:

- *Economic benefits.* FRS offers higher transmission speeds, simplified network design, and low operating costs compared to earlier packet services. FRS could also be viewed as an economic service, because it provides a migration path to ATM.
- *Availability.* FRS is widely available and is currently supported by IXCs, LECs, and RBOCs. FRS is also increasingly supported by the AAPs.

While FRS can provide universities with these benefits, frame relay technology has several shortcomings, including lack of support for isochronous traffic, complex administration, bandwidth limitations, inconsistent product offerings, transmission delays, and packet loss.

4.5.4 ISDN

ISDN provides universities with another data and video communications alternative. This interconnection approach involves the use of switched *digital* facilities to link a student with a training site located across the city or the country. ISDN provides $n \times 64$ Kbps end-to-end digital connectivity with access to voice and data services over the same digital transmission and switching facilities. The more well-known interfaces are 2B+D, which provides two switched 64-Kbps channels, plus a 16-Kbps packet/signaling channel (144 Kbps total); and 23B+D, which provides universities with DS1 speeds.

ISDN is now seeing a renewed push toward deployment. ISDN can support the following IDL applications:

- *Access to the Internet.* RBOCs are increasingly addressing the needs of universities for access to the Internet through ISDN solutions.
- *Access to groupware.* ISDN can support both public and private groupware solutions.
- *Two-way compressed videoconferencing solution.* The RBOCs are developing this capability to meet the needs of universities.

ISDN provides universities with the following benefits:

- *"Wideband" bandwidth.* A group of ISDN solutions are available that can support the bandwidth needs of universities for lower-end videoconferencing solutions (128 Kbps–1.544 Mbps).
- *Availability.* The RBOCs have been actively deploying ISDN to support the networking needs of universities, as well as other applications.

4.6 SUMMARY

A number of IDL solutions are available to support the university IDL applications. Table 4.1 correlates the IDL applications and solutions, while Table 4.2 correlates the IDL application solutions with the networking solutions.

Table 4.1
University Distance Learning Application Solutions

Application Solutions	K–12 Applications				
	Collaborative Learning	*Access to Databases*	*Access to Libraries*	*Remote Education*	*Access to Supercomputers*
Internet	×	×	×	×	×
Groupware	×	×	×		
One-way video/two-way audio		×		×	
Two-way audio/video	×			×	

Table 4.2
University Distance Learning Networking Solutions

Application Solutions	Networking Solutions			
	ATM	*SMDS*	*FRS*	*ISDN*
Internet	×	×	×	×
Groupware	×	×	×	×
One-way video/two-way audio	×			×
Two-way audio/video	×			×

References

[1]   Applebome, P., "Colleges Luring Students With Discounts in Tuition," *New York Times*, Vol. CXLIV, No. 49,921, Dec. 25, 1994, p. 1.
[2]   National Research Network Review Committee, Computer Science and Technology Board, "Toward a National Research Network." Commission on Physical Sciences, Mathematics,

and Resources National Research Council. Washington, D.C.: National Academy Press, 1988, p. 14.

[3]   "Industry and Academia Partner in Pioneering 2-Way Video Distance Learning Service," *The Heller Report*, Dec. 1993, p. 4.

Bibliography

Hindin, E., "ATM Deployment: Notes From the Leading Edge," *Data Communications*, Sept. 21, 1994, pp. 90–91.

Martin, B. L., "Using Distance Education to Teach Instructional Design to Preservice Teachers," *Educational Technology*, Vol. XXXIV, No. 3, March 1994, p. 51.

Rohde, D., " SMDS, Switched LANs Win Battle for School Net," *Network World*, Nov. 21, 1994.

State-Based Distance Learning Initiatives

5

5.1 OVERVIEW

As discussed in several places in this book, many of the distance learning initiatives originate at the state level. This chapter provides a synopsis of many of those initiatives as they were ongoing at press time. This discussion is not intended to be exhaustive; more activities than listed herewith are actually underway in many, if not all, states. The material is based in part on an analysis by Hezel Associates, which is updated yearly [1]. Interested readers should secure the latest copy of this document for detailed state-by-state analysis, which goes well beyond the snapshot provided here.

In addition to discussing the initiatives themselves, this material furnishes a sample of the gamut of technologies that currently are being employed to support IDL. By and large, these technologies are state of he art, although they tend to fall in the wideband rather than in the broadband (i.e., ATM) area. Many of these initiatives are multiyear undertakings. It is expected that a significant share of the initiatives described here will continue to be of importance for years to come.

5.2 ALABAMA

In 1993 Alabama courts voiced concern that Alabama operated an educational system that favored students living in affluent areas. The state has been ordered to reformat its education-funding mechanisms to fix those inequities and to find other appropriate remedies. The comprehensive reform bill Alabama First: A Plan for Academic Excellence Act of 1994 includes an education technology trust fund for programs that focuses on use of technology in education. The proposed fund would be used for video distance networks, a statewide computer network, and a two-way messaging system. Alabama currently has three networks that support education-related programming:

- The *Alabama Public Television* (APT) system reaches 97% of the population. APT delivers two feeds of instructional programming during the academic year. In addition, it broadcasts Integrated Science 6, 7, and 8 programs to almost 50,000 middle school students each year through SERC (discussed in Chapter 1).
- SDENET is an educational computing and communications system linking the main office of each school district to the state *Department of Education* (DOE). It also includes a bulletin board for educators. The network is funded by the state; in addition, the software and the training for the system are provided without charge.
- The Alabama Supercomputer Network is a system of DS1 communication lines reaching all research universities in the state, as well as 20 high schools. It is partially used for scientific computing. The network is operated by the state. Users also have access to other educational institutions and state agencies free of charge.

Other university-sponsored DS1-based interactive compressed-video systems are also employed.

5.3 ALASKA

The state of Alaska has been working on a plan to achieve cohesiveness among telecommunications users across the state, under the auspices of *Telecommunications Information Council* (TIC). Many state agencies are involved, including the state DOE, the Division of Information Services, the University of Alaska, the Alaska Public Broadcast Commission, and the state legislature. TIC was launched in 1992 to develop strategies for the integration of telecommunications in the state and has focused its attention on the Satellite Interconnection Project; it has also addressed the issue of instructional and technical needs within the state. A number of telecommunications forums have convened in recent years directed toward building a consensus among potential users. However, the state's ability to carry through with implementation of large-scale telecommunications systems has been retarded by fluctuations in oil prices, which are the financial basis of the state budget.

There has been recent legislative activity related to distance learning. House Bill 106 proposed to establish the Alaska Education Technology Program in the DOE to (1) provide technical assistance and training to schools and libraries; (2) plan for coordinating and expanding existing networks; (3) investigate the development of new networks for educational usage; and (4) establish he Alaska Education Technology Fund to provide a percentage of the money needed to support distance learning (specifically, to purchase, install, and maintain technology in classrooms and libraries, provide training, and provide access to networks). House Bill 107 proposed the issuance of bonds for the purchase of classroom equipment and library computer automation. Alaska currently has two networks that support education-related programming:

- The *University of Alaska Computer Network* (UACN) interconnects school districts across the state and supports access to the Alaska Teleconferencing Network, commercial services, and the Internet.
- The DOE utilizes the *Rural Alaska Television Network* (RATNET) satellite for overnight feeding of instructional TV materials.

Other university-sponsored satellite-based one-way video systems are also employed.

5.4 ARIZONA

Issues of interest to IDL are addressed by the *Arizona Educational and Informational Telecommunications Cooperative* (AEITC), a consortium of public universities, community colleges, K–12 schools, and the Department of Administration. AEITC has four objectives: (1) to coordinate the development of a state telecommunications plan that serves current and evolving needs; (2) to assist members, including the state, in assessing their telecommunications needs and to recommend solutions; (3) to encourage deployment of statewide telecommunication networks; and (4) to hold open forums for information gathering and dissemination. Arizona currently has several networks that support education-related programming:

- EDLINK allows educators modem access over an 800 number to e-mail services, the Internet, and bulletin-board service. The service, managed by Arizona DOE, has been used primarily in K–12 schools.
- The *Arizona State Public Information Network* (ASPIN), jointly operated by Arizona State University, Northern Arizona University, and the University of Arizona, is a statewide network of users that includes K–12 schools districts, community colleges, universities, hospitals, libraries, and industry.
- NAUNet, managed by Northern Arizona University, is an interactive voice, data, and video network connecting the campus with electronic classrooms at various community college sites, university sites, high school sites, and a state capital site in Phoenix shared by the Arizona DOE

and the Arizona Supreme Court. Recent expansions include the addition of 20 community colleges and high schools sites.

- AzTeC Computing is a network of local users in the Tempe area managed by the Arizona Telecommunications Community Computing Network. The network involves the cooperation of key universities in Arizona.

5.5 ARKANSAS

In the past couple of years, distance learning in Arkansas has grown significantly. Several new entities have appeared, including the *Arkansas Public School Computer Network* (APSCN), IMPAC Learning Systems, Inc., *Arkansas Educational Television Network* (AETN), and the Arkansas State Technology Planning Group. AETN recently installed a satellite uplink and has sponsored numerous telecourses and conferences. All colleges and universities in Arkansas now have access to Internet services. In 1993, the Public Service Commission ruled favorably on an SBC (formerly Southwestern Bell) infrastructure development plan. SBC agreed to direct excess earnings for five years toward building a DS1/DS3 network in support of state telecommunications.

The Arkansas State Technology Planning Group, an ad hoc committee overseen by the governor's office and staffed by the Department of Computer Services, is chartered to develop a state telecommunications plan. However, Arkansas continues to rely on out-of-state resources for some of its distance learning needs (e.g., in recent years there has been a slight increase in the number of nationally originated telecourses and instructional TV for students in the state). Another initiative involves the choosing of 40 schools by IMPAC each year from the state's 269 school districts to participate in a program of computer hardware and software installation and upgrading.

In terms of IDL networks, SBC works closely with the Department of Computer Services in defining communication technologies that can be used by the state, including microwave, satellite, and fiber optics. AETN reaches over 90% of the state; local cable companies also carry those programs. APSCN is the state's designated K–12 instructional and administrative computing network. Several colleges also operate and maintain Ark-Net, a computer network established in the early 1990s.

5.6 CALIFORNIA

Planning for distance learning in California is quite visible, as observed from statewide initiatives involving broad-based educational entities. Aggressive and progressive recommendations made in the California Master Plan for Educational Technology place the state in the forefront of IDL planning. Specifically, the deployment of a statewide integrated telecommunications network for education, called the Golden State Education Network, was one of key recommendations of the Master Plan, released in 1992. The Master Plan aims at improvement of existing educational and informational resources as well as the establishment of new resources. In the early 1990s, Pacific Bell launched the "Education Initiative," which promises to allocate $100 million to connect, by 1996, all California libraries, K–12 schools, and community college campuses in its service territory, using ISDN for audio, video, and data transmission.

The California State University has been studying what would be the appropriate telecommunications infrastructure necessary to support information systems and technology-related programs. The university has 20 campuses throughout the state. The inclusion of technology-related costs in all capital projects for new buildings and campus renovation is now routine, as recommended in the document *Telecommunications Infrastructure Planning Guidelines*, distributed to campuses. California now has the following statewide and local IDL networks:

- The *Learning Solutions Network* (LSN) is a computer-based network established by California State University that also reaches 30 public high schools. LSN distributes coursework developed at the University of Illinois in Urbana.
- The University of California's *Intercampus Telecommunications Network* (ITN) connects nine campuses to the San Diego Supercomputer Center and supports e-mail and library services.
- The *Educational Telecommunications Network* (ETN) is a satellite-based network that broadcasts live, interactive, staff-development programs to school districts statewide. ETN also produces and delivers information to health care professionals, hospital workers, and vocational students. ETN is planning a move to digital compression technologies to decrease costs, by reducing required transmission capacity, and to increase security.
- The *Bay Area Regional Research Network* (BARRNET), which operates in northern California, and the *California Education and Research Foundation Network* (CERFNET), which operates in southern California, are two regional research networks funded by the NSF. They connect university research centers, private colleges, and community colleges.
- The California State University operates CSUNet, a university-wide communications network that links several community colleges. CSUNet supports voice, data, and videoconferencing services.
- Pacific Bell has established the Knowledge Network Gateway, which allows users to access university libraries, databases, e-mail, computer bulletinboards, and the Internet. This enables students and faculty to communicate with friends and colleagues in other states and other countries.

- The TECHNET Network is a California State University network that connects faculty and students to businesspeople to facilitate real-world participation.

5.7 COLORADO

Coordinated IDL planning in Colorado is achieved under the auspices of the Colorado *Telecommunications Advisory Commission* (TAC), which has representatives from the telecommunications industry, state industry, state government, and K–12 and post-secondary education. The state Department of Administration and Departments of Education and Higher Education are two key government agencies that participate in TAC. IDL activities are undertaken based on the following dynamics: (a) a mandate from the legislature to reduce educational inequalities in rural areas; (b) a need to provide health care in rural areas; (c) the passage of Amendment One in 1992, which reduces the amount of funding available for education; (d) the desire to stimulate the state economy by aiding in training and economic development; (e) the need to support under-served (economically and geographically) disadvantaged students; and (f) the requirement of the Colorado Commission of Higher Education for teacher recertification.

The Northeast Alliance of *Colorado Learning Network* (CLN) has requested funding from the TIIAP/NTIA (see Chapter 1) to implement a plan to interconnect IDL providers across the state, in order to achieve more affordable and quality education. To expand service, the network is slated to connect educational organizations; educational/training partners; businesses; industry; municipal, country, and state government agencies; health care providers; libraries; and museums. CLN is also working with the Colorado Division of Telecommunications for digital upgrades to a backbone network capable of supporting videoconferencing. The following major education-oriented networks serve the state:

- The Department of Administration operates a microwave network for state agency use. There is a five-year plan to upgrade the system to provide bandwidth to support transmitting video to state users, including educational institutions.
- Higher-education institutions are linked through various networks and communication technologies. The University of Colorado system manages a fiber network to link its campuses. The University of Northern Colorado uses leased DS1 lines to deliver compressed video. Various community colleges are developing networks that use either fiber or ITFS distribution methods. Colorado SuperNet supports access to the Internet and also links Colorado state libraries.

5.8 CONNECTICUT

During the early and mid-1990s, various IDL initiatives have been pursued in the state. Connecticut Public Television manages an ITFS that is key to the IDL activities in the state; however, educational institutions are also independently planning for IDL. Furthermore, the state DOE has begun an effort aimed at up-graded technology in public schools. Higher education bodies, such as state community and technical colleges, the Connecticut State University system, and the University of Connecticut, have pursued IDL programs. The *Joint Committee on Educational Technology* (JCET) has submitted plans to the legislature that include long-term goals and objectives and a series of recommendations for action. The following networks are of relevance:

- Knowledge Net is a statewide ITFS network owned by the DOE and operated by Connecticut Public Broadcasting, Inc. Knowledge Net supports educational, government, and business applications.
- The community and technical college system of Connecticut operates a statewide cable network that links 16 cable systems via an ITFS channel.

5.9 DELAWARE

The state of Delaware has been slow in supporting IDL initiatives. As a result, advances in that area have been limited. Some pilots have been conducted (e.g., PolyTech High School in Woodside), with possible future expansion (e.g., the Woodside network may be expanded to link three schools and those members of industry that contributed to the project). The College of Nursing of the University of Delaware system now offers a graduate level course via satellite links to nursing students at three sites; other departments are expected to offer degree programs. The new Delaware statewide DS1 network is used to meet the networking needs of state government employees, including members of the state DOE, institutions of higher education, and libraries. A basic service that is available is integrated e-mail.

5.10 FLORIDA

Planning for the integration of technology in Florida schools is a task of the DOE, which is attempting to purchase a transponder on the Telstar 401 satellite through state funding. Programming would be delivered via VSAT earth stations at schools and other locations; digital video compression is expected to be successfully employed.

Practically, Florida IDL activities center on the publicly funded *Florida Information Resource Network* (FIRN), which is the information highway that connects school districts, community colleges, and universities in the state: All 67 school districts, 28 community colleges, and 9 state universities are linked through FIRN. FIRN continues to add connections of schools and community colleges to the Internet on an ongoing basis. FIRN, established by the Florida DOE in 1982, is a network of dialup and leased lines based on the existing state university system's network. The network is funded by the legislature and is utilized for both administrative and instructional purposes. FIRN allows electronic transfer of student data and records. In the early 1990s, FIRNMAIL, an e-mail service, was inaugurated (more than 50% of FIRNMAIL users are in K–12 schools). The network also supports an electronic conferencing function, as well as access to full-text databases and classroom resources.

Through the use of terrestrial DS1 lines and satellite services, Florida's DOE is planning several electronic classrooms located at a number of state agencies. This video capability will enable the agencies to meet electronically, as well as present instructional and informational programming to their staff and clients.

Another resource is the *Florida Remote Learning Services* (FRLS), which provides a statewide IDL network. FRLS has managed a consensus process among diverse interests, so that interoperability between systems can be achieved. The Florida Public Broadcasting Service also has many educational outreach services (one-way video, two-way audio programming); these are sent by satellite to over 5,000 teachers statewide. There are also over 200 ITFS channels licensed to 17 educational institutions, counties, and cooperatives in Florida.

SUNSTAR is a satellite network linking a number of universities, vocational schools, K–12 school districts, and community colleges in the state. The network utilizes steerable C/Ku-band satellites to create a statewide teleconferencing capability for education, government, and private industry. Video, voice, and data are also provided by the State Division of Communication's SUNCOM network (FIRN utilizes SUNCOM for communication service, which has negotiated special rates for all its user agencies.) The Division of Communication is also looking at ISDN and ATM for possible use.

5.11 GEORGIA

IDL activity in Georgia in recent years has centered around funding issues. The Distance Learning and Telemedicine Act of 1992 (Senate Bill 144) and a state lottery, started in 1993, are the key funding mechanisms. Under this act, the Georgia Department of Administrative Services was directed to establish networks for IDL and medicine applications. The state lottery has funded satellite and computer initiatives. Over 750 downlinks for schools, libraries, and regional education centers were secured; in addition, K–5 schools in the state have received computers. Specifically, the *Georgia Public Telecommunications Commission* (GPTC) has purchased two transponders on Telstar 401. In addition, the lottery funds are being used to support technology training centers and to purchase IDL programming.

GPTC delivers public broadcasting and instructional TV to K–12 and higher education institutions (the DOE selects, purchases, and schedules daytime public TV programming for secondary and elementary schools). The *Georgia Statewide Academic and Medical System* (GSAMS) is a statewide network of over 100 sites linked by DS1 lines. In 1994, the network connected more than 300 sites with two-way video capability; these sites included schools, adult and technical learning centers, and universities and colleges. As another example, PeachNet is a data network connecting 34 state-funded institutions; it is used for administrative, instructional, research, and business services. Though PeachNet, the *Georgia College Educators' Network* (GC EduNET) provides e-mail, conferencing, file sharing, and online databases to educators and educational administrators.

5.12 HAWAII

Providing equal access to educational opportunities throughout the state of Hawaii is a challenge because of the geographic dispersion of the population. In Hawaii, both lower and higher public education are directed on a statewide basis: A single DOE provides K–12 instruction and a single *University of Hawaii* (UH) system with 10 campuses and 5 education centers. The Hawaii Educational Networking Consortium develops and promotes the use of networking to fully interconnect the Hawaii education and research community to provide IDL services, including access to libraries and information resources.

Hawaii has two statewide networks: the *Hawaii Interactive Television System* (HITS) and the *Hawaii Wide Area Integrated Information Access Network* (HAWAIIAN). HITS provides interisland video services for the University of Hawaii, the DOE, state agencies, and others. Analog microwave carries video signals from Oahu to other islands, with local ITFS distribution. HAWAIIAN consists of an interisland SONET system. All cable TV companies can receive HITS. As part of the cable TV franchise agreements, cable TV companies must also provide support for institutional networks in the state. These institutional networks interconnect schools, campuses, and governmental buildings. They utilize various technologies including SONET and Ethernet over cable.

Maui Community College developed and operates Skybridge, an analog point-to-point microwave system, which supports two-way interactive video transmission between sites on Maui, Molokai, and Lanai.

5.13 IDAHO

In the early 1990s, the governor of Idaho significantly restructured the state's *Telecommunications Council* (TELCOM), which, for a number of years, has guided the development and evolution of the state's telecommunications infrastructure, including the *Idaho Educational Public Broadcasting System* (IEPBS)/Higher Education Microwave Network. Now TELCOM's emphasis is more focused on policy development. Ongoing initiatives relate to increasing backbone capacity, funded via state legislature appropriations. Recently there was also the creation of the Idaho Technical Advisory Committee, which includes legislators and representatives from the DOE, the *Department of Transportation* (DOT), and other government agencies. The mandate of the department is to plan and coordinate telecommunications activities in the state.

IEPBS operates a microwave network linking PBS stations and translators scattered throughout the state. IEPBS also supports the IEPBS/Higher Education Microwave Network, which provides courses in engineering, teacher training, primary health care for nursing, pharmacology, and physician/primary care training.

5.14 ILLINOIS

In the late 1980s and early 1990s, the State of Illinois Department of Central Management Services began the implementation of a network that carries voice, data, and video services to nodes in 20 Illinois cities. The agency provides IDL and videoconferencing through its satellite uplink; coincidentally, this allows access and transmission of data to a national area. This state network has joined with universities and state agencies providing students, faculty, and state employees access to the National Research and Education Network. By 1994, the state network linked Springfield, Peoria, and Chicago and supported data, voice, and video. IDL connections have been implemented at Eastern Illinios University, in Charleston; Western Illinios University, in Macomb; and Black Hawk Community College, in East Moline. Recently Governor's Educational Technology Task Force was formed, which is chartered to prepare a strategic plan for the use of technology in education. Millions of dollars have already been allotted to acquire video equipment.

5.15 INDIANA

Ameritech, Inc., and the state of Indiana have signed a deal to create what they claim is "the world's largest distance learning network," which will link up to

1,700 Indiana schools to the state's information superhighway. Between 1994 and 1999, Ameritech also will invest approximately $120 million in infrastructure to extend advanced communications services to interested schools and other locations in its Indiana service territory. Under a previously approved alternative regulatory plan called Opportunity Indiana, Ameritech will make provisions to link schools and others. The Indiana Utility Regulatory Commission approved Opportunity Indiana in 1994. Under the regulatory plan, Ameritech-Indiana will reduce basic local residential service rates by $2.21 per month over the next two years and cap them until 1998.

The Corporation for Educational Communication (CEC) is an independent not-for-profit corporation that provides grant administration, project planning, and administration for programs that enhance the quality and availability of education in Indiana through the use of new communications technologies.

CEC chose Sony equipment to use in the distance-learning network being developed in Indiana. Funded with an initial $30-million grant from Ameritech Corp., Indiana's distance-learning project is administered by CEC. Indiana schools have the opportunity to go online by creating consortia based on geographic location. The consortia prepare distance-learning proposals for the CEC, which evaluates them and awards funding based on merit and diversity. More than 17 Indiana schools have gone online, with estimates that 500 distance-learning sites will be installed over the next six years. Other grants include $300,000 for the state's Access Indiana program to help stimulate a statewide information system for Indiana residents, businesses, and institutions, and $50,000 to the PSI Foundation for Homework Hot Line, an educational voicemail application.

The Indiana University (IU) School of Continuing Studies has added 20 VTEL MediaConferencing and two multipoint bridging systems to their extensive distance-learning network. The Indiana University's Multicampus Technology Project (MCIP) has made it possible for 10 campuses to offer a variety of credit and noncredit educational programs to Indiana citizens wherever they live, work, or study. The videoconferencing classrooms were the first to go online with a DS1 network that includes advanced technology auditoriums and broadcast studios. The project is also experimenting with running the VTEL systems over an ATM network to conserve costs.

Parkervision, Inc., announced that its CameraMan System II Instructor Camera System will be utilized to provide distance-learning classroom equipment for up to an estimated 600 to 700 Indiana middle and high schools over the next six years.

As the demand for programming expanded, the Indiana Higher Education Telecommunications System (IHETS) began looking at more efficient methods of signal delivery to keep pace with the growing requirement. With this expanded mission in mind, it became obvious that satellite communications could provide a viable solution. Currently there are 55 receive sites connected to the IHETS network via Radiation Systems' 1.8-meter receive-only VSATs and integrated receiver decoders. The system provides one-way video and two-way audio with plans calling for online fax and computer interface. The network originally operated over a combination of DS3 fiber-optic leased lines and ITFS terrestrial microwave radio. It provided closed-circuit TV instruction to the state's high schools and universities. Restrictions associated with the leased lines/ITFS microwave radio solution included the recurring cost of the DS3s, the area of coverage for the line-of-sight microwave system, and the lack of available frequencies for the ITFS to provide more channels and more programming. In January 1992, a 7m Ku-band uplink and headend supplied by Radiation Systems of Duluth, Georgia, was installed. It was put at the combined campus of Indiana University and Purdue University in Indianapolis. Sites other than schools and universities are also benefiting from the IHETS network. Users include county extension offices, hospitals, state agencies, and private businesses.

5.16 IOWA

Iowa is a state known for its advances in IDL. Iowa has a statewide fiber optics network, the *Iowa Communications Network* (ICN), which connects over 100 educational endpoints in all counties, three universities, and the Iowa Public Television network. ICN is a statewide, 3,000-mile fiber optics network providing video, voice, and data services to public and private educational institutions and state agencies. ICN brings in the equivalent of 12 DS3 links to each of the 99 county points of presence; construction of the network began in the early fall of 1991 and was ready for use in the fall of 1993. Educational endpoints include K–12 school districts, community colleges, universities, area education agencies, and private colleges. Additional transmitter sites, private colleges and universities, and prison locations are being added.

ICN scheduled its first 15,000 hours of programming in 1993. Over 35,000 hours of instruction per semester were being contemplated. However, there is a movement afoot for the commercialization of the network. ICN was designed to be implemented in three phases, two of which are in place. Phase III would extend services to all high schools, selected libraries, and additional state agency sites. However, appropriate funding for that part of the network remains an issue. Individual schools are able to move forward with Phase III at this time, but at their own expense. Most schools have been waiting until the legislature takes action before making a decision about their connections. (ICN is discussed further in Chapter 6.)

5.17 KANSAS

The Kansas Board of Regents recently created the Systemwide Telecommunications Task Force to examine IDL opportunities. The task force provides recommendations that identify current and future initiatives at Regents institutions that use telecommunications technologies and describe how those telecommunications initiatives will be funded. In addition, the task force was charged with developing a vision, including goals and objectives, for using IDL in the Regents system.

KANS-A-N is the state government's digital telecommunications network, supporting voice, data, and video communication throughout the state. It is used by all government entities in the state, including 26 of the state's compressed-video sites. The University of Kansas Medical Center uses KANS-A-N as a backbone network for telemedicine and medical education via compressed videoconferencing. KANS-A-N is used by higher education to deliver IDL courses to sites in Kansas and Missouri (this last being achieved via a link between the Kansas and Missouri systems). The *Kansas Regents Network* (KARENET) connects Regents institution computers to one another and to KANS-A-N.

The Regents Educational Communications Center is a teleproduction facility at Kansas State University that provides fixed and mobile satellite uplink capabilities (compressed video) with downlink sites more than 125 locations within the state (most of these sites are located in public schools or educational service centers).

Two systems deliver K–12 IDL instruction in Kansas. One is a compressed video of 25 sites statewide; the second is a state fiber-optic system that delivers full-motion analog signals. No formal planning exists to connect these systems, but the analog clusters contain codecs that make them compatible with the compressed-video sites.

5.18 KENTUCKY

All the commonwealth's universities, the Council on Higher Education, the Kentucky Department of Information Systems, and Kentucky Educational Television are developing the *Kentucky Telelinking Network* (KTLN). KTLN is a two-way DS1 network supporting compressed video, audio, and data. Among other goals, KTLN is expected to: (a) improve access to education, health and human services, and economic development resources; (b) foster an environment in which Kentucky's universities and government agencies can leverage resources by sharing a multipurpose system for the information and service delivery; and (c) support the Kentucky Education Reform Act (K–12). KTLN has its roots in the existing regional networks operated by Murray State University the University of Kentucky, the University of Louisville and Western Kentucky University. Sites will be developed at each of the eight recognized technology "innovation centers" and "pilot centers," school districts established in suppor of the Kentucky Education Reform Act. The KTLN expansion would add be tween 28 and 40 additional sites. Existing regional networks used to deliver IDL programming are approaching maximum utilization. These university net works are as follows:

- Murray State University transmits a complete degree program to three re mote sites in its service area.
- The University of Kentucky offers a majority of courses required for degrees in physical therapy, clinical laboratory science, and Master's level nursing.
- Western Kentucky University delivers several dozen courses each semes ter to its extended-campuses centers in Glasgow and Elizabethtown.

The *Kentucky Education Technology System* (KETS) has two infrastruc ture components: the *Education Communication Network* (ECN) and the *Education Information System* (EIS). The ECN is a fiber-based network linking regional service centers. The EIS consists of the administrative and instruc tional software supporting KETS. All 176 school districts were expected to be connected by 1995.

Kentucky Educational Television (KET) provides a number of video and data systems. KET's public broadcast system consists of 15 public TV stations and six translators. The two-channel satellite delivery system provides instruc tional video materials, live interactive high school courses, professional devel opment seminars, and teleconferences. (KET would like to employ digital compression for its satellite system, which allows an increase from its current 2 channels to 10 programming channels.) In addition, KET provides KETNet, a computer network that enables schools within the state to communicate until fiber lines are in place.

The West Kentucky *Interactive Telecommunications Network* (ITN) is a land-based two-way distance learning and economic development system. ITN now connects two public schools and three community colleges, as well as the University of Louisville's Telecommunications Research Center. The network satisfies the state's focus to share, not duplicate, resources by delivering high school, community college, and four-year college courses.

5.19 LOUISIANA

Louisiana Public Broadcasting (LPB) provides a statewide TV network for edu cation. It has a six-station network that is carried by more than 150 cable TV systems in the state. The system migrated from microwave to digitally compressed satellite delivery in 1994. A transponder on Telestar 401 is employed to deliver the programming (the transponder also allows for at least two more digitally compressed educational channels to be delivered to schools across the state and across the United States).

The *Louisiana Instructional Satellite and Telecommunications Network* (LISTN) provides satellite services to higher-education institutions in northern Louisiana. Northwestern State University is also planning a switchable network using ISDN and DS1 lines. In addition, the Office of Rural Development has a number initiatives, all related to rural issues, including the following:

- The Nursing Program. Four sites are connected via DS1 links to hospitals. Nurses take IDL classes over the DS1 lines supporting video, voice, and data.
- The Satellite Communication Initiative. Community health education is delivered to rural areas by TCI and other cable TV companies.

5.20 MAINE

In recent years, Maine has progressed from planning to implementation of a statewide IDL network. A funding priority has been to expand Maine's Education Network to high schools, college, and universities in the state. The goal is to provide two-way interactive video to all locations.

The Education Network of Maine uses fiber trunks to connect the seven University of Maine campuses with two-way, full-motion video. On each campus, courses are distributed to multiple sites via point-to-point microwave and ITFS channels.

5.21 MARYLAND

Maryland's Information Technology Board was created in 1993 by executive order to serve the needs of the DOE, higher education, and all state government agencies. The goal of the educational technology initiative is to establish a full-motion two-way video IDL system for the state. Four pilots in Baltimore City high schools and three in western Maryland community colleges were underway at press time. The network uses a SONET ring backbone (OC-48) and provides three DS3s' (OC-3/155 Mbps) worth of bandwidth into each classroom, allowing for the transmission of commercial-quality video as well as data and multimedia. The architecture of the network allows it to migrate to ATM. By 1996, about 700 educational entities were expected to participate, eventually reaching up to a total of 1,800 sites. Maryland Public Television will become a key video and data services provider to educational sites. The Maryland Center for Public Broadcasting currently operates a six-station statewide public TV system to deliver K–12 and college courses.

METNET, the Maryland Educational Technology Network, is part of the National Learning Link Consortium. It is a computer-based service that provides free e-mail, bulletinboards, file transfer, conferencing, and forums to K–12 educators and other interested individuals throughout the state.

5.22 MASSACHUSETTS

The *Massachusetts Corporation for Educational Telecommunications* (MCET) has a central role in IDL activities in the state: the state legislature has selected MCET to develop and implement a statewide data and voice infrastructure. MCET partners include the Massachusetts Education Computer Network, Massachusetts Board of Library Commissioners, the *Center for Educational Leadership and Technology* (CELT), Moakley Center at Bridgewater State College, the University of Massachusetts and Five College Partnership, the Merrimack Education Center, DOE, Bolt Beranek and Newman (BBN), and PI-net.

During 1993, CELT was awarded a contract to develop a statewide technology plan. CELT has advocated a client/server network with nodes in each sizable community. By 1994 work had begun on obtaining funding for putting the plan into action. That plan is expected to support the state's recently enacted school-reform law. Cooperation among higher education, K–12 education, library, and municipal communities is part of the design that will result in many community networks linked into a statewide infrastructure.

A "virtual" magnet school is currently being created in the Merrimack Valley of north-central Massachusetts. The magnet will lead and challenge students from 14 high schools throughout Massachusetts by exposing them to current research problems and methods. The program utilizes teaching and research faculties from the University of Massachusetts. A number of other initiatives were underway at press time, and a number of IDL networks were available.

5.23 MICHIGAN

Michigan has experienced a crisis in the financing of education in recent years because a 1993 bill eliminated property taxes as a basis for the support of education (in 1994, Michigan voters approved a 2% increase in the sales tax to fund education at a statewide minimum level of $5,000 per student.) Nevertheless, funding for IDL is proceeding, supported by a Public Service Commission order that requires telephone companies to return excess earnings to a special fund for education. For example, as of 1994, Michigan Bell's excess earnings, with interest, amounted to $12 million; the PSC also required Michigan Bell to match the excess earnings, generating a fund of about $24 million. The Michigan Council on Telecommunications Services for Public Education has been put in charge of reviewing all telecommunications projects, specifically the expenditures from the ratepayers' portion and Michigan Bell's matching portion. The executive order from the governor, however, raised objections from parties such as cable TV operators and long distance carriers, who felt that Michigan Bell would benefit excessively. As a result, a second executive order was issued that ensures that expenditures of the fund are not limited to a particular transmission technology but could include cable TV, Ku-band satellite, and so on.

One network, MICHNET, is a statewide data network in nine of Michigan's publicly supported universities. MICHNET manages and operates the *National Science Foundation Network* (NSFNET) in cooperation with corporate partners. Dial-in numbers in 23 Michigan cities allow computers with modems to access the network through local telephone calls. The *Michigan Information Technology Network* (MITN) operates two satellite networks: Business Network, serving the business sector, and EdNet for K–12 schools. Business Network has 12 sites statewide, and EdNet has 38 downlink sites serving 59 schools. The DOE was the first Michigan agency and one of the first education departments to develop a Gopher node on the Internet for access to information about education in the state.

5.24 MINNESOTA

In the early 1990s, the legislature authorized the purchase or lease of IDL equipment in K–12 schools. A council was also constituted with the following tasks: (a) to develop a statewide vision and plans for the use of IDL technologies; (b) to develop educational policy relating to telecommunications; and (c) to oversee coordination with campuses, K–12 education, and regional educational communications. In particular, there has been a lot of support for statewide K–12 access to the Internet.

Through the auspices of the *Higher Education Coordinating Board* (HECB), Minnesota is in the process of implementing a network for compressed video, funded by the state legislature. Eventually the network will provide links to 10 hub sites throughout the state. In parallel, *Minnesota Equal Access Network Services, Inc.* (MEANS) is developing the state's first commercial videoconferencing network, called MedNet. This fiber-based network, which can support existing analog systems, has the capability of transmitting two-way video to colleges and universities, hospitals, government agencies, and businesses throughout the state. There are plans to interconnect MedNet to all educational TV networks and state-run government networks.

MnSAT is a satellite network financed by the technical college system; it utilizes a Ku-band as well as a C-band uplink, providing access to government agencies, business, and industry. The MnSAT network utilizes a number of transmission technologies, including analog and digital video over microwave, coaxial cable, fiber optics, and satellite.

5.25 MISSISSIPPI

Mississippi's Legislative Committee for Statewide Telecommunications has the responsibility of researching the statewide telecommunications systems. The committee works with existing agencies to establish guidelines for interconnectivity and to ensure accessibility. Because of several setbacks, the state's ITFS network was still in the planning stages at press time. Management of the system rests with EdNet, a public board that also includes representatives from state agencies. Recently, EdNet and its member agencies were granted construction permits for channels at 11 locations in the state, giving EdNet authority for a total of 220 ITFS channels. An additional set of construction applications was pending.

The DOE is in the process of implementing the *Mississippi Online Network Exchange* (MONEX). MONEX is a data exchange computer system that connects all school districts in the state with the DOE. Applications include data base access, electronic fund transfer, and other administration functions.

5.26 MISSOURI

In 1993, the state legislature approved the creation of a 15-member Commission on Information Technology. The commission is charged with (a) developing a state telecommunications strategy by using technology and facilities available to all residents; (b) enhancing and equalizing educational opportunities; and (c) providing greater access to information (health care delivery and economic development opportunities also are priorities.) The commission was to report its findings and present its recommendations to the general assembly and the governor by 1995.

The Outstanding Schools Act authorized a program of grants to public school districts for the use of technology. The specifics are in the development stage under the counsel of the *Video Instructional Development and Educational Opportunity* (VIDEO) advisory committee. Grants under the act were to be awarded for the 1994–1995 academic year.

In Missouri, there are two major IDL networks. The University of Missouri system implemented a DS1 backbone network with microwave links connecting its four campuses and the state capital for voice, data, and compressed-video services (the system interconnects with Kansas's KANS-A-N). MOREnet, the Missouri Research and Educational Network, is a statewide data network established in the late 1980s. The goal of the network is to develop, maintain, and foster applications for electronic-based education, research, and service to its members. All public colleges and universities, private higher-education institutions, and K–12 schools can access the network.

Additionally, ShareNet is a collaborative partnership that comprises school districts, university systems, business, industry, and other community partners in Kansas and Missouri in the Kansas City metropolitan area. The purpose of the network is to advance teaching in this service area through technology-enhanced instruction. Mention also should be made of three interactive TV clusters that are providing 20 courses to 600 students at 13 schools through an analog fiber and copper system. In addition, 400 schools districts participate in ITFS programs.

5.27 MONTANA

The *Montana Educational Network* (METNET) is the state's IDL network. METNET supports two-way compressed video to more than six major sites (university and community college campuses). In addition, METNET has switched video access to any Sprint user or Sprint public meeting room. Although further METNET implementation was on hold at press time due to financial constraints, different partner agencies are expected to individually pursue other telecommunications initiatives. METNET also includes dailup *Bulletin Board Service* (BBS). The BBS is a 15-node system that links K–12 schools and universities in a statewide informational network, enabling the exchange of ideas among teachers, administrators, and students. In addition, the Department of Administration (DOA) was planning to serve as a testbed for Internet access in the state.

Another facility, the Big Sky Telegraph, is a free dail-in system that links rural schools to information resources; it supports mentor relationships among teachers and provides a mechanism for community development projects. The program has received funding from the Annenburg/CPB Math and Science Project and the U S WEST Foundation for supporting math and science through IDL methods. Recently, a $2.5-million initiative to help rural elementary and secondary educators learn how telecomputing can be of help was launched. The project is aimed at ending the isolation of rural math and science teachers: This rural telecomputing initiative will give teachers the technical skills, hands-on experience, and ongoing support to incorporate the resources of the Internet and other computer networks into their own classrooms and curriculum.

5.28 NEBRASKA

Recently Nebraska increased the scope of IDL and satellite programs through the implementation of a new telecomputing network. Some observers view the state's multiple and diverse educational telecommunications systems as a model for statewide coordination of telecommunications. Few states have succeeded as well as Nebraska has in offering access to broadcast, ITFS, satellite, microwave, and fiber optic networks through one agency, the *Nebraska Educational Telecommunications Commission* (NETC). The NETC also sponsors the *Agricultural Satellite Corporation* (AG*SAT), discussed in Chapter 1. The *Collaborative State Plan for Technology in Nebraska Schools* enunciates five priorities: collaborative implementation of the state plan, dissemination of technological information, establishment of training service guidelines, development of a compatible infrastructure, and universal student access.

As an IDL network, the Nebraska ETV Network offers a variety of services through its nine TV stations and dispersed translators. The University of Nebraska at Lincoln Television provides instructional and public TV programs along with production services. The University of Nebraska at Omaha Television also produces programs. The DOE, in cooperation with NETC, offers elementary and secondary instructional TV programming through the Schools TeleLearning Service. ETV services also include the *Hearing Impaired Video Information Service* (HI-VIS) for hearing-impaired viewers, a *Descriptive Video Service* (DVS) for sight-impaired residents, and a closed-captioning production capability. In addition, ETV provides EduCable, a continuing education cable TV service, through a dedicated channel provided by Omaha and Lincoln cable TV systems.

The NEB*SAT system uses two full-motion video channels for public TV and instructional services, a dozen compressed-video channels, and public radio distribution. There are plans to also provide for a fiber optic network service to complement the satellite system.

5.29 NEVADA

Higher education is a major focus behind IDL planning in Nevada. With only two universities, located at opposite ends of the state, there is a need to make programming available to Nevada residents. The University of Nevada system has several IDL initiatives. Two universitywide computer networks exist between the University of Nevada at Las Vegas and the University of Nevada at Reno. Compressed video has been used for instruction, beginning in 1993. Some courses are offered using an audiographic equipment transmitting voice, data, and graphics. The university and community college system operates NevadaNet, a DS1-based network reaching 13 locations, including associated campuses. The system also operates a microwave network; community-funded facilities distribute educational programs to remote locations.

5.30 NEW HAMPSHIRE

Limited funding has restricted the growth of New Hampshire's IDL initiatives. IDL activity is concentrated in the state's university system. Specifically, the University System of New Hampshire operates an IDL network with sites at the University of New Hampshire-Durham, the University of New Hampshire-Manchester, Plymouth State College, and Keene State College. The network uses DS1 compressed, two-way video; each site on the system is capable of both receiving and originating programs. The system has also been used to deliver training to state agency employees; interactive PBS Adult Learning Satellite Service conferences and workshops are also broadcast. As another effort, in the mid-1990s, a Governor's Task Force on Telecommunications, compriseing members from education and telecommunications industries, was organized to provide guidance for the development of NHNet, a statewide computer network.

5.31 NEW JERSEY

The New Jersey DOE's recent efforts have been directed at linking constituents and organizations while maintaining existing staffing levels in the face of government cutbacks and monetary freezes. The State Plan for Educational Technology Task Force, created by the DOE, released in 1993 the report *Educational Technology in New Jersey: A Plan for Action*. Representatives from school districts, higher education, business and industry, research laboratories, museums, libraries, and government and community agencies participated in the process. The outcome of this five-year plan included efficient communication for schools and comprehensive access to and use of technology in appropriate areas of education.

New Jersey Link is a collaborative effort between the DOE and New Jersey Public Broadcasting. It is a computer network that provides electronic mail, discussion forums, and databases to over 3,500 teachers, administrators, parents, board members, and postsecondary faculty throughout the state.

The *New Jersey Intercampus Network* (NJIN), established in the mid-1980s, is a network of higher-education users. NJIN's efforts have included recommendations for both intercampus data and video communications protocols. The result of its activities are the *John von Neuman Computer Network* (JvNCnet) for data and the *New Jersey Network* (NJN) for video. Approximately 25 colleges and universities are linked to the Internet via JvNCnet for electronic communication and access to remote library and computer resources.

NJIN has a three-phase plan for the development of an intercampus video to augment the receive-only satellite links existing on many New Jersey campuses. The first phase, linking William Patterson College to Richard Stockton College by utilizing a microwave backbone, was already complete at press time. Phase 2 will extend the microwave network to other colleges and universities in critical areas of the state. Phase 3 will add channels and institutions to the system.

New Jersey Public Broadcasting uses broadcast, ITFS, satellite, cable TV, and fiber optic technologies to offer instruction throughout the state. Each semester, about 20 college telecourses are available from more than two dozen participating institutions in New Jersey, Delaware, New York, and Pennsylvania. In addition, NJN delivers over 100 K–12 instructional programs each year. Furthermore, the Cable Television Network of New Jersey, Inc., a statewide system reaching more than 1.7 million households, offers 30 hours of for-credit programming from several institutions of higher education each week.

5.32 NEW MEXICO

The Educational Technology Coordinating Council, assembled in the early 1990s by the state Board of Education and the Commission on Higher Education, advises those two entities on policies to facilitate education through the use of technology. Of specific interest, the council coordinates IDL efforts among participants.

In recent years, higher-education entities have increased their use of educational telecommunications. For example, the University of New Mexico transmits engineering, medical, and professional training. Another example is the New Mexico Educators' *Network for Educational Communication* (NEDCOMM), which is a free educational data communications network.

The *Manufacturing Engineering Project Network* (METNET) operates New Mexico, Inc., a liaison for the state's manufacturing sites and educational institutions. The goal is to facilitate requests from manufacturing entities for training from higher education institutions. New Mexico Inc. is a clearinghouse for courses from many sources.

The State Board of New Mexico delivers legal education to lawyers and others via live interactive workshops. The program is accessible, throughout the United States by home satellite receivers.

5.33 NEW YORK

New York's Telecommunications Exchange brings policymakers together with industry representatives, telecommunications users, and other interested parties, with the goals of developing and recommending a strategy for the state aimed of establishing a viable infrastructure to support evolving communication applications, including IDL. Recently, they recommended that the state foster an open environment that will encourage competition, preserve universal access to affordable, high-quality service, and bring a new gamut of information capabilities to businesses and consumers.

The *Technology Network Ties* (TNT) network links more than 400 school districts, all *Board of Cooperative Educational Services* (BOCES) units, libraries, other educational institutions, and the *State Education Department* (SED). The TNT network consists of computers, communication facilities, and telecommunications hardware and software that allow resource sharing and enhanced access to information and services.

The *State University of New York Satellite Network* (SUNYSAT) began operation 1989. All 64 *State University of New York* (SUNY) campuses have been equipped with satellite antennas to receive courses, professional training, and teleconferences for schools, public and private colleges, state agencies, and prisons. SUNYSAT is looking to create four digital channels from one of its analog channels, for combined data and video transmissions. In a related initiative, the New York Network provides a microwave network for the state's 10 public broadcasting stations to relay instructional TV programming from the state capital.

SUNYNET, a digital voice and data network providing a telephone and computer link among SUNY campuses, will interconnect the state's community colleges. Through a series of gateways, SUNYNET connects with TNT, the *New York State Educational and Research Network* (NYSERNET), and the *Higher Education Services Commission* (HESC). NYSERNET is a research computer network serving educational institutions, government agencies, libraries, hospitals, and other industries involved in education and research. NYSERNET affiliates include more than 200 sites throughout the state.

Four of New York's community-licensed public broadcast stations participate in LearningLink, a not-for-profit computer network for educators. LearningLink, originating at WNET in New York City, is the largest public broadcast station in the country. (LearningLink now operates out of PBS as a national consortium of 22 sites.)

5.34 NORTH CAROLINA

North Carolina has one of the most ambitious coordinated statewide IDL plans in the country. In 1993 the Governor's office and the state's largest telephone companies—BellSouth, Carolina Telephone Co., and GTE South—announced plans for a statewide IDL network known as the *North Carolina Information Highway* (NCIH), which is expected to link 3,300 locations in the state by the year 2004. The state legislature allocated $4.4 million to support installation of an initial 104 sites (which was accomplished in 1994); those initial funds were allocated to pay for technology at schools, medical centers, and universities across the state. Additional funding will occur throughout the project lifecycle.

The NCIH utilizes new technologies such as fiber optic cables and ATM switches, as well as traditional technologies such as telephones, computers, and TV sets. The network allows for access to such resources as video teleconferencing and library retrieval. The network will connect corporations, universities, colleges, community colleges, public schools, hospitals, and eventually homes. The establishment of the NCIH requires the education institutions in the state to redefine their telecommunications plans to fit into the statewide NCIH vision. In many cases, existing telecommunications networks will integrate into the new network. Organizations that so far have acted independently will be required to collaborate.

By 1995, 13 geographically dispersed community colleges in the state were connected with the NCIH, and 11 more will be linked soon thereafter. In addition, the state Department of Community Colleges has established a satellite-based statewide telecommunications system (called EDNET) connecting its 58 campuses. The C/Ku-band receive antennas provide educational programs, occupational planning, literacy enhancement, and economic development.

The NCIH will employ ATM switches interconnected by SONET facilities. Under an arrangement with the state, the telephone companies will build, maintain, and own the physical network. As noted, existing subnetworks are now in the process of connecting with NCIH. For example, the CONCERT network, a data and two-way high-quality video network supporting services for several universities and research institutions, will be consolidated into the NCIH. Operated by the Center for Communications, a consortium of 16 universities, research institutions, and graduate centers, the DS3-based CONCERT network connects sites across the state and delivers interactive video to more than 50 facilities.

The North Carolina Agency for Public Telecommunication operates two satellite-based networks. The Open Public Events Network connects more than three million homes with state government officials and services through satellite and cable TV. The State Services Network provides teleconferences and education to over 250 *TV receive only* (TVRO) satellite receive antennas in schools, community colleges, medical centers, and other public buildings across the state.

5.35 NORTH DAKOTA

The *North Dakota Educational Telecommunication Council* (NDETC) is the key telecommunications coordinating organization in the state. Recently the coun-

:il undertook a needs analysis, along with an assessment of available technolo-
;ies. The goal was to propose strategies and implementation plans for IDL-
)ased education. The Governor's Telecommunication Task Force has begun
:onsidering the telecommunications needs of the state across a variety of state
ιgencies. The task force includes members from education, state agencies, busi-
1ess, and community organizations. In addition, the *Center for Innovation in
'nstruction* (CII), which includes representatives of the Department of Public
nformation, vocational education, and the *North Dakota University System*
NDUS), provide technology support and training to statewide K–12 schools.

North Dakota has a gamut of networks that incorporate satellite, micro-
ʌave, ITFS, fiber optics, telephone lines, cable, and broadcast TV. The princi-
)al IDL network is the North Dakota *Interactive Video Network* (IVN), operated
)y NDUS and the *Information Services Division* (ISD). The network involves a
;eries of first-tier and second-tier hubs capable of supporting audio, voice, and
lata signals. Under the *Educational Telecommunications Council* (ETC) plan,
VN will become the backbone video network to which all schools will eventu-
ιlly connect. Intercluster connectivity is based on two-way analog video (in-
eroperability with an evolving digital backbone must still be worked out).

NDUS and ISD cooperate in the operation of a number of other informa-
ion and data networks. Specifically, the *North Dakota Information Network*
NDIN) is the main network operated by ISD, of which several other networks
ιre part. SENDIT, funded by NDETC, now a statewide network, serves as the
lata access network, also connecting the statewide library network. Further-
nore, Prairie Public Television recently started the Prairie Satellite Network,
ʌhich connects 70 sites, mostly schools, throughout North Dakota. The net-
ʌork links seven broadcast stations; as of press time, it has an uplink and 70
atellite downlinks antennas.

;.36 OHIO

)hio's IDL activities are guided by Project Equity, which is the state's technol-
•gy program that redresses identified inequities in K–12 schools. For example,
n 1995 the Project funded 25 activities. Also resulting from Project Equity was
. study related to K–12 teachers' receptivity toward video technology for class-
·oom instruction. Telecommunications policy is under review by the Ohio
)OE, the State Board of Education, and the Ohio Public Utilities Commission,
ʌith the goal to develop coherent policies that complement each other and pro-
·ide maximum assistance to Ohio schools.

The *Ohio Educational Broadcasting Network Commission* (OEBNC) oper-
tes a two-way microwave and satellite network to deliver educational material
⊃ K–12 schools (public and private). A number of hospitals are also connected
⊃ the network using microwave links. A number of postsecondary schools also

rely on this network for access and programming. The *State of Ohio Networ*
for Integrated Communications (SONIC) provides connectivity to statewid
agencies using an extensive microwave network. The network supports all stat
telephone services and data communications. The Ohio Data Network trans
mits information to various state departments via the SONIC network.

The *Ohio Academic Resource Network* (OARnet) aims at facilitating dis
semination of knowledge throughout the state. This is accomplished throug
connection to a number of national and international networks. Academic in
stitutions in Ohio are encouraged to connect to OARnet. The system connect
the supercomputer at Ohio State University by utilizing the SONIC network
More fundamentally, most, if not all, schools in Ohio have access to compute
services and the Internet through the DOE and the *Ohio Educational Compute*
Network (OECN). In addition, the National Public Telecomputing Network op
erates the Cleveland Free-Net, which offers a number of services for schools
such as bulletinboards, e-mail, and information access.

5.37 OKLAHOMA

In the early 1990s, there was a vote in favor of a state bond issue that include
$14 million for IDL. At the same time, the Division of Telecommunications i
the Office of State Finance and the Board of Regents for Higher Education pre
pared two plans for an integrated statewide telecommunications system fc
education and government. One plan provides online services to state agencie
and education institutions; the other plan provides integrated voice, data, an
video telecommunications to state agencies, schools, and colleges. The plan
include options for satellite and fiber optic transmission facilities and are de
signed to overcome geographic coverage limitations and to eliminate redur
dancy in the state's telecommunications systems. The fruition of these plans i
far from complete. There has been support for the development and use of th
Oklahoma Government Telecommunications Network (OGTN). This would i
turn provide free access to a DS1 digital network by all K–12 schools and acces
to DS3-based facilities in the late 1990s.

In more recent years, the Oklahoma DOE has actively sought funding fc
telecommunications projects, with varying degrees of success. Through th
Board of Regents, ONENET, the Oklahoma WAN with a connection to the Ir
ternet, became a reality in 1993. This project also placed many of the commu
nity libraries online.

H.B. 2133 created the State Data Processing and Telecommunication Ac
visory Committee, which advises the director of the Office of State Financ
The committee consists of representatives of the state DOE, the Regents, th
Department of Libraries, *Oklahoma Educational Television Authority* (OETA
the Department of Vocational and Technical Education, and a number of sta

departments. The Advisory Committee and its subcommittees establish criteria for the selection of telecommunications projects to be funded

A number of statewide and local IDL networks are available in the Oklahoma. For example, Talkback Television connects the state's public and private colleges and universities, community colleges, and technical institutions, as well as businesses and medical facilities. The system utilizes fiber optics, microwave, ITFS, and cable TV systems for distribution of courses to 64 sites statewide. OETA, which operates the state's public TV system, shares its microwave system with Oklahoma State University and Talkback Television. Since the early 1990s, OETA has provided a broadcast channel called The Literacy Channel in Oklahoma City. In addition, the DOE operates Specialnet, which is a computer network for K–12 schools. The system allows the DOE Service Centers, vocational-technical agencies, and other agencies to communicate and share information.

5.38 OREGON

Oregon was one of the first states to create an agency specifically dealing with IDL. The Ed-Net Board approved the design of a three-network system. Network I delivers one-way video, two-way audio instruction at broadcast video quality via satellite to 200 subscribing schools, education services district, libraries, medical centers, state agencies, and government locations. Additional state agency sites have been added over time. Network II is a digital satellite network piloted in 1992. It includes over three dozen sites at state colleges, universities, community colleges, and other locations. Network II supports two-way compressed video (up to 15 channels can be delivered simultaneously). There are plans to extend Network II to at least 40 sites. Network III, the COMPASS computer network, uses dialup service to interconnect schools, colleges, libraries, and state agencies and has more than 2,000 users.

5.39 PENNSYLVANIA

IDL planning in Pennsylvania is distributed among various local, regional, and institutional groups. A movement toward more coordinated planning is now more visible. Act 67 of 1993 encourages accelerated private investment in broadband telecommunications networks. There have also been legislative discussions to create a fund to support loans to schools for the leasing IDL equipment.

A number of IDL-based networks now service the state. Some school districts have satellite and/or computer networks in place. The University of Pittsburgh is working with the Pittsburgh Public Schools to link all the schools in the district over a computer network. The Learning Link is available throughout the state and is coordinated by the Pennsylvania Public Television Network. A freenet has also been put in place.

The governor's office announced in the early 1990s that the commonwealth would make available its network, PANET, to all the 500-plus school districts and other educational institutions, at costs considerably less than they were then paying for communication services. (PANET links all commonwealth agencies and offices statewide; it also provides the backbone of the state's high-speed computer communications system that connects the state's research universities.)

Pennsylvania State University connects 17 campuses via satellite, and 4 campuses via terrestrial DS1-based links; these links carry compressed video. A two-way video microwave system connects the University Park campus of Pennsylvania State University with the Hershey Medical Center, the capital campus at Harrisburg, and Behrend College; the system is used for the dissemination of undergraduate and graduate level instruction. The *Pennsylvania Public Television Network* (PPTN) represents the nine public TV stations in arranging and distributing funding and programming. The public TV system consists of two-way microwave links. In addition, the DOE operates PENN-LINK, a computer network for data, e-mail, and bulletinboard services for secondary schools and other educational entities.

5.40 RHODE ISLAND

There has been an increasing level of K–12 activity in recent years in Rhode Island. This can be attributed to the Public Utilities Commission's decision to allow NYNEX to waive installation and access fees for schools and libraries. By 1994, 1,500 lines had been installed, which has resulted in increased Internet access. RI-NET, the network supporting schools' access to the Internet, consists of four regional gateways: Brown University, the University of Rhode Island, the Naval Underwater Sea Command at Newport, and the Providence-based public TV station, WSBE. All automated library systems are linked by the Library Board of Rhode Island's Library Telecommunications Network. Interconnect, a consortium of seven local cable TV systems, offers educational instruction from the Community College of Rhode Island on its public access channels.

5.41 SOUTH CAROLINA

In South Carolina, a state Video Users Advisory Council, created by legislation in the early 1990s, assesses needs and assists in setting policy on resource allo cation, consumption, and funding issues. The council is composed of representatives from the Office of Information Technology, the state Budget and Control Board, Higher Education, the state Department of Education, the Commission for Higher Education, the state Department of Corrections, the Criminal Justice Academy, the Health and Environmental Control Board, community colleges, and *South Carolina Educational Television* (SCETV).

Recently, SCETV implemented the Multichannel Digital Satellite Network, a 20-channel digital video system delivering information to about 250 sites. The existing ITFS system, which serves K–12 education, was planned to be incorporated into this satellite network. IDL service will be expanded to include higher education institutions, state agencies, and other uses. The network will provide recieve sites with courses and programs from national network providers, for example, international news feeds from 43 countries in 20 different languages. SCETV also produces programming for the SERC (see Chapter 1). With SCETV, business sites will receive teleconferences and training programs originating from the Department of Health and Environmental Control, the Employment Security Commission, the state Development Board, the governor's office, and several technical colleges. SCETV transmits more than 110 hours of programming each day (over the multiple channels). SCETV offers medical education, legal continuing education, law enforcement training, state agency training, school staff development, and early childhood education. The SCETV system has five production facilities and 11 transmitters. Each TV and FM radio station in the system is connected by two-way microwave and by satellite. SCEV uses ITFS for last-mile delivery to 350 middle and high schools. One hundred or so schools that are not fully served by ITFS receive a satellite dish and two digital recievers (for simultaneous channel reception). The *National Instructional Satellite Service* (NISS) uses SCETV as its distribution point. Twenty-five regional distribution centers act as tape and delay feed centers for ITFS system programming and serve as production studios for short-distance education courses for school districts.

5.42 SOUTH DAKOTA

South Dakota has the third-largest population of adult learners participating in IDL. South Dakota's 15 site *Rural Development Telecommunications Network* (RDTN) uses terrestrial lines and satellite to deliver compressed video to the state. RDTN is the result of recommendations from a 1991 Governor's Telecommunications Task Force meeting. All public universities and many of the technical institutes participate in the network. The RDTN includes 15 studios, 6 universities, 3 postsecondary technical institutes, and 3 hospitals. The backbone of the network is a DS3 network leased by the state's telecommunications division from U S WEST. Government grants facilitated the acquisition of additional DS1 lines, video-compression equipment, and studio equipment. Higher-education institutions are becoming more involved with RDTN, although most still offer their own IDL courses. About half of RDTN's two-way video sites are located at the six state universities and three of the four postsecondary technical institutions.

TECH NET, a satellite system operated by the South Dakota Office of Adult, Vocational and Technical Education, has 12 downlinks located at vocational education sites throughout the state. One-way video programs are delivered to the receive sites. Another resource is the *Regents Information System* (RIS), which is a statewide data communications system operated by the Board of Regents supporting e-mail for students, faculty, and staff. *South Dakota Public Broadcasting* (SDPB), operates interconnected networks of eight TV stations and nine radio stations. SDPB recently acquired three downlinks for delivering programs to schools and universities.

5.43 TENNESSEE

Tennessee has a 10-year master plan, called *FYI Tennessee*, aimed at the statewide modernization and deployment of technologies, including advancement of IDL. In recent years, the state's IDL activity has exhibited a larger coordinated effort among key participants. IDL players include the state Board of Regents and Higher Education Commision, DOE, the Tennessee Valley Authority, and the Public Service Commission. The Tennessee Board of Regents Ad Hoc Distance Education Committee submitted its report in 1993, which recommended the following: (a) the establishment of a permanent IDL committee, (b) a technology committee at each campus, (c) the endorsement of the *Tennessee Board of Regents System Distance Education and Telecommunication System Plan* and (d) protocols for interconnection between IDL institutions and other organizations.

In the early to mid-1990s, Tennessee has migrated from a simple point-to point network architecture to a statewide system. The *FYI Tennessee* plan has led to the deployment of intelligent network capabilities and high-speed fiber optic connections in government offices. The plan calls for ISDN links in all 95 counties by the end of 1998 and ATM services by the year 2000.

TEN is a statewide effort to provide each school district with administrative software; it is the product of a joint plan involving the Office for Information Services, the state DOE, and the governor's office. Utilizing the state's Electronic Tandem Network, TEN links K–12 with universities; it will support professional communication statewide and worldwide. TEN will also transport administrative data. Now TEN is addressing itself to data transfer, and instructional support, as well as access to the Internet to every K–12 school in the state. The state Electronic Tandem Network is planning pilot ATM: Th network architecture will migrate from its initial low speed and DS1 digital services to ATM, as business needs dictate (some dedicated DS3 links are already in use).

The Tennessee Board of Regents upgraded their TECNET Data Network to support many Internet services. The network was recently expanded to 46 sites. Another network, the Tennessee Elk River Development Agency, provides Internet access for communities through the Elk River system. In addition, Elk River offers community access for IDL and telemedicine; it uses traditional compressed-video and DS1 lines.

5.44 TEXAS

Progressive legislative support allows Texas to maintain a leadership position in IDL. IDL has benefited from state government actions related to access and to remedies for inequities. Specifically, there has been the establishment of an education tariff for telecommunications services in support of IDL. H.B. 653 grants qualified educational institutions reduced rates (equal to 75% of the usual rate) to meet the IDL needs of the state. In addition, the Texas Interactive Multimedia Communications Fund Demonstration Program (H.B. 1029) established funds for school districts to acquire multimedia technology and services. Initiative H.B. 183 encourages cooperation among education, government, and private industry in demonstration projects related to IDL.

The *Texas Education Network* (TENET) is a K–12 communication network established in the early 1990s as a dial-access system. It provides e-mail, bulletinboard, computer conferencing, and database query services to more than 30,000 users. TENET links the Texas Education Agency, Regional Education Service centers, professional organizations, other state agencies, and all school districts.

Texas School Telecommunications Access Resource (T-STAR) is an initiative undertaken by the Texas Education Agency to provide an integrated telecommunications system of one-way video, two-way audio, and data services using both TVRO and VSAT antennas. About 36 school districts received a TVRO system in Phase I. In 1994, the system was used for K–12 programming and two-way videoconferencing (using both DS1 and satellite links). Phase II of the project, including VSAT implementation, began in 1993. The plan is to have every public school district in the state interconnected by the end of 1997.

State of Texas Administrative Resource Line (STARLINK) is a satellite-based teleconference network established in the late 1980s by the Texas Higher Education Coordinating Board. The network connects many community and technical colleges in the state. STARLINK uses interactive videoconferencing for the professional development of educators statewide. For example, in 1994,

STARLINK produced or distributed teleconferences to more than 9,000 participants.

5.45 UTAH

The Utah Education Network Consortia establish policies that encourage the cooperative use of IDL in the state's higher education and K–12 systems. It includes executive-level administrators from higher-education institutions and employees of the DOE. The Utah *Education Network* (EDNET) is an interactive, two-way video, microwave TV system connecting 35 sites across Utah and providing voice and data channels to colleges, universities, and public schools. EDNET has been expanding to incorporate a compressed-video network for data and video transmissions. Participation in EDNET at the K–12 and postsecondary levels is increasing. More than 120 high schools and junior high schools have been added in the recent past. EDNET is a "multiple option telecommunications network," providing users with alternatives such as (a) public TV stations; (b) Utah's Learning Channel (a full-power VHF broadcast TV station licensed to the Utah State Board of Regents and operated by the University of Utah for the state's colleges, universities, and public education institutions); (c) public radio stations; (d) interactive, two-way video, microwave TV systems; (e) ITFS transmission (in the Salt Lake City area); (f) satellite distribution and reception; and (g) Teletext via the state public TV network's vertical blanking interval.

5.46 VERMONT

IDL activity in Vermont still remains somewhat disjointed; there is recognition that coordination of technology among institutions is necessary to maximize resources. In recent years, the *Vermont Education Technology Consortium* (VETC) provided funding for Internet access. The Statewide Systematic Initiative in Science, Mathematics, and Technology has helped schools connect to a telecomputing network. Vermont is now developing a statewide telecomputing network, VTNet, linking schools, businesses, and higher education.

Vermont's *Distance Learning Strategic Plan* suggested, in the past couple of years, the formation of an educational telecommunications planning leadership structure for the state. Such a lead agency would encourage cooperation and communication among the planning entities in state agencies, education, and private industry now individually developing IDL systems. The *Vermont Interactive Television* (VIT) system is a DS1-based, two-way audio and video system that uses compressed video. VIT uses its own multipoint contro-

switching system, which allows each site to switch and exchange e-mail to and from more than 12 sites.

5.47 VIRGINIA

In spite of budget limitations, Virginia remains a leader in IDL planning and infrastructure development. Virginia has increased the number of services available to education. Old Dominion University and the Virginia Community College System manage TeleTechnet, a statewide satellite network. Eighty tele-courses leading to six baccalaureate programs were offered during the 1994–1995 academic year. The *Department of Information Technology* (DIT) operates a statewide system supporting audio teleconferencing for education entities and agencies. Access facilities connect six satellite uplinks at the University of Virginia, Virginia Tech, Old Domain University, WCVE-TV Richmond, and community colleges.

The University of Virginia operates a network, VRNET, that includes a computer bulletinboard for student teachers and their professors. VCCNet is a computer network that is operated by the Community College System of Virginia. The *Virginia Satellite Education Network* (VSEN) is a statewide K–12 satellite network administered by the DOE. In addition, the DOE also operates *Virginia's Public Education Network* (VA.PEN), a computer network serving over 10,000 users in schools and higher education.

5.48 WASHINGTON

The state of Washington is proceeding with initiatives in IDL in several areas. The Washington Department of Information Services has developed and implemented the *Strategic Plan for Video Telecommunications* and the *Strategic Information Technology Plan*. The *Washington Interactive Television System* (WITS) provides a broadcast-quality TV studio, postproduction services, satellite services, cable channel coordination, and two-way interactive videoconferencing. *Washington State University* (WSU), Educational Service District 101, and the Community and Technical College System of Washington have collaborated with the Department of Information Services to develop a joint telecommunications planning document, called the *Video Telecommunications Strategic Plan*. The document identifies common needs in the state and provides recommendations for meeting them through telecommunications technologies. The charter of the plan is to implement, by 1997, a shared statewide video telecommunications system integrating new technologies and existing resources to serve government, education, and the public.

The Washington Department of Information Services was charged by the legislature with the task of ensuring the cost-effective development of a shared video telecommunications system to serve the legislature, state agencies, public schools, educational service districts, community and technical colleges, universities, state and local governments, and the public. Along those lines, *Washington Interactive Television* (WIT) developed, through partnerships, a TV studio with satellite broadcasting, downlink coordination, cable distribution, and two-way videoconferencing to 13 sites statewide with dialup access.

The *Satellite Telecommunications Educational Programming* (STEP) network, which operates from Educational Service District 101 in Spokane, delivers programming to the northeast. The state of Washington has more than 150 STEP sites. STEP uses a one-way video, two-way audio system to reach high school students and teachers in 14 states. The *Washington School Information Processing Cooperative* (WSIPC), a joint venture of nine Educational Service Districts, is currently installing a telecommunications network that will be used for data and voice transmission and, eventually, the delivery of video between sites. *Washington State University* (WSU) houses the *Washington Higher Education Telecommunication System* (WHETS), an interactive video and audio instructional system. WHETS recently expanded to a three-channel digital and one-channel analog system, using DS1-level compressed video.

5.49 WEST VIRGINIA

In West Virginia, the *Distance Education Oversight Committee* (DEOC) advises and recommends the planning, funding, and implementation of distance education initiatives in West Virginia. DEOC is a joint committee established by the Board of Directors of the State College System and the Board of Trustees of the University System. The West Virginia *Distance Learning Coordinating Council* (DLCC), which has statutory responsibility to plan and deploy the state's IDL network, has the following goals: (1) to expand access to IDL education by providing downlinks for all public secondary schools; (2) to increase the receive capabilities of higher-education institutions by adding second channel receive capability; (3) to enhance the public's receive capability by equipping libraries and other public facilities with satellite antennas; (4) to establish new origination sites; and (5) to perform other related functions.

West Virginia is incorporating a gamut of technologies, including DS1 links, microwave links, fiber links, and satellite links to serve the state's telecommunications needs. ISDN is being deployed, and frame relay is being tested as a probable forerunner to ATM. The State College and University System of West Virginia has planned, built, equipped, staffed, and authorized a satellite telecommunications system, SATNET, which is designed to deliver instruction that supplements campus courses. While the initial focus of SATNET programming has been at the graduate level, a demand for undergraduate courses has developed. SATNET currently has 24 higher education receive sites. The West Virginia Educational Information System (WVEIS), a data network operated by the West Virginia DOE, offers public educators computing capabilities. ETN is a statewide DS1 multiservice digital backbone operated by the Department of Administration, which all state agencies are able to use. Higher education uses the ETN backbone for the *West Virginia Network for Educational Telecompting* (WVNET). WVNET is a data center supporting administrative, educational, and research projects in higher education statewide and serves as an entry point for the Internet. Three other digital DS1 data networks are also carried on the state ETN backbone: WVEIS, the DOE's Statewide Administration Network; BS/CE, the DOE's statewide K–6 network; and the Administration system.

The West Virginia *Educational Broadcasting Authority* (EBA) operates a microwave transmission system that supports programming among the state's public TV stations. EBA connects colleges and universities to the higher-education uplink; EBA also feeds K–12 TV programming to local cable companies that serve public schools. Also overseen by the state's DOE, the *West Virginia Microcomputer Educational Network* (WVMEN) is a dialin network. E-mail, bulletinboards, public domain software, and conferences are offered. There are plans to expand the system to include connections to the Internet.

5.50 WISCONSIN

The *Distance Education Technologies Initiative Committee* (DETIC) was formed in 1992 under the direction of the state of Wisconsin *Educational Communications Board* (ECB). Soon thereafter, DETIC completed a statewide IDL study that included needs assessment and analyzed available and projected technologies for meeting identified needs.

In another study, the Governor's Blue Ribbon Telecommunications Infrastructure Task Force presented a competitive approach to expanding the economic and social benefits of IDL. Telecommunications were identified as a key strategic focus for several issues, including education. The study recommended ways in which the state can (a) manage the transition to a competitive communications marketplace, (b) remove barriers to competition and effective use of telecommunications, and (c) stimulate private sector development of an enhanced telecommunications infrastructure.

Continuing Education Extension (CEE), a Division of the University of Wisconsin-Extension, developed the *Distance Education Implementation Plan*, which is being carried out. IDL funding has been allocated by CEE to encourage program and course development.

The ECB is the licensee of five of the state's eight public TV stations. It also operates a statewide radio service for the delivery of instructional programming on its subsidiary FM subcarrier channel. In addition, ECB coordinates 17 regional ITFS systems and sponsors Learning Link, an online e-mail and bulletinboard system. State funding has been provided, through the ECB, to two fiber optic–based IDL networks. The first network connects seven K–12 school districts in northeast Wisconsin and the other links three K–12 districts, the regional Cooperative Educational Service Agency, and five campuses of a technical college in northwest Wisconsin. A third fiber optic system was funded by three K–12 school districts and is coordinated by a technical college in central Wisconsin.

WISCNET is a statewide network serving as a gateway to the Internet for the University of Wisconsin system, technical and private colleges, and state agencies. WISENET is a computer bulletinboard and e-mail system for K–12.

The University of Wisconsin-Extension Bulletin Board, an online bulletinboard and e-mail system, is used by the a number of universities to exchange information about teleconferences. Also, a community "freenet" has been created by the University of Wisconsin-Milwaukee.

5.51 WYOMING

The Wyoming State Telecommunications Video Network has been implemented. Also, the *University of Wyoming* (UW) has developed a DS1-compressed-video network that uses fiber optic lines of the *Wyoming Statewide Network* (WSN) to deliver data, voice, and video.

Reference

[1]   Hezel Associates, *Educational Telecommunications, State By State Analysis*, Syracuse, NY, 1994.

Part II
The NII and Distance Learning

Part I explored the distance learning needs of K–12 schools, universities, and corporations. It also described potential application and networking solutions to support those requirements. The solutions, however, are not ubiquitously available, nor are they universally affordable. Some argue that ubiquity and affordability cannot be accomplished without the establishment of a national information infrastructure (NII). The creation of the NII is driven by a number of stakeholders: (a) state and federal government agencies; (b) the traditional telephone companies, including the Interexchange carriers and the LECs; (c) the emerging network service providers, including the cable TV companies, and the AAPs; and (d) the Internet community. The goal of this section of the book is to explore the role of each of these groups in creating the NII and to identify to what extent those efforts are adequate to support distance learning applications.

Part II consists of four chapters. Chapter 6 discusses the current and potential roles of state and local governments in creating the NII and in supporting the expansion of distance learning programs. Chapter 7 is devoted to the current and potential initiatives of the telephone companies related to the NII and to distance learning. Chapter 8 focuses on the emerging roles of the (interrelated) cable TV and AAP industries. Chapter 9 examines the Internet initiatives, which are a key element of NII and of distance learning.

NII and Distance Learning Initiatives of the Federal and State Governments

6

The role of the federal and state governments in the United States in creating the NII and in promoting education is a subject of intense debate among interested parties. Stakeholders have different views regarding the government's role. Those views are shaped by the stakeholders' own self-interests as well as by their political beliefs, which range from the far right, which advocates a minimum role for government, to the far left, which advocates a more extensive government role. The real debate, however, ought to focus on what the nature of that role should be. Whatever government's role is, it should not increase the budget deficit (and therefore the national debt) or raise taxes.

To explore the role of the government in creating the NII and supporting IDL, this chapter first defines the concept of NII. It then explores the current NII/IDL initiatives of state governments and the federal government and the strengths and limitations of those activities. The chapter concludes by identifying the possible changes in the directions of government policies regarding the NII and IDL.

6.1 DEFINITION OF THE NII

In nonlegal terms, the NII is the aggregate of all information technology required to create a universal, affordable, integrated, seamless, interactive, and flexible digital communications infrastructure in the United States that facilitates unimpeded any-to-any connectivity for any commercial, industrial, educational, and governmental purpose. Many do not see the NII as one network; instead, they see it as adding up to multiple systems. In other words, they see the NII as "a network of networks." These multiple networks have been created to (a) serve the diverse needs of a variety of vertical industries (education, health, science, business, news, and entertainment) and (b) serve the particular interests of the individual groups of stakeholders involved in the effort.

One the most debated issues in recent times has been: What is the benefit of an NII? The Clinton-Gore administration views the NII as a tool to create more jobs, to educate the public, and to create a more equitable environment, by eliminating (or at least reducing) the difference between the information haves and have-nots. Politically conservative thinkers have different views regarding the benefits of the NII. They disagree with the administration regarding the importance of the NII in creating jobs, citing the examples of the traditional telephone companies, which have been laying off people while making major contributions toward the creation of the NII [1]. They do, however, agree with the administration about the important role of the NII in improving the productivity of the American educational system through the creation of IDL and multimedia training systems. The *National Technological University* (NTU), the largest electronic university in the United States, offering the programs of 40 universities, sees a great role of the NII in supporting IDL: "The NII will clearly support the new service needs of schools, museums, libraries, hospitals, colleges, and universities" [1].

State and local governments have been playing active roles in supporting the creation of an NII. Their roles are shaped by history, the political orientation of each administration, changing political climates (by state and over time), the challenges facing the governments, and institutional goals.

6.2 NII AND DISTANCE LEARNING INITIATIVES OF STATE GOVERNMENTS

State government share the following goals with respect to NII and IDL:

- To enhance the competitive position of the state relative to the other states;
- To educate citizens in rural areas, to close the information gap between them and those in "more-informed" suburban communities;
- To increase the number of jobs in the state.

To accomplish those goals, state governments play the following roles relative to the creation of the NII:

- An end-user role;
- A regulatory role;
- An educator role.

Each of these roles is examined next, followed by proposed changes or additions to those roles.

6.2.1 The End-User Role of State Governments

State governments throughout the United States have been financing the creation of statewide information networks, as discussed in detail in Chapter 5. Those networks are increasingly forming the nucleus of the NII. One example of these statewide networks, the NCIH, was described in Chapters 2 and 5 and is further discussed in Chapter 7; NCIH is of particular interest because it used broadband technologies to support state-of-the-art communication. Another example of a statewide network is the *Iowa Communications Network* (ICN). The following is a description of the state of Iowa's experience in building the ICN and a comparison between the ICN and the NCIH.

6.2.1.1 Objectives of the ICN

The creation of the ICN was driven by the state of Iowa's desire to link all the K–12 classrooms to form a "virtual schoolhouse." The network is intended to (a) enable K–12 students to access courses unavailable in their local schools and (b) enable schools and libraries to share resources.

ICN is a 3,000-mile backbone network that was completed in late 1993 and that interconnects all counties in Iowa. Institutions interconnected in the network include state universities, 15 community colleges, 2 private colleges, and 54 high schools.

The network is based on a SONET platform. A fiber optic drop at DS3 speeds is provided to each participant in the ICN. Classrooms are configured as follows. The instructor's TV monitor outputs the audio and video of each remote classroom in a fixed sequence, one at a time. The sequence is interrupted when a student uses a push-to-talk microphone to ask the teacher a question. This arrangement is different from the NCIH arrangement, which provides each classroom with continuous presence throughout the duration of the IDL session.

The Iowa government owns and operates the network. Unlike the NCIH, the state has build a parallel network to those built by the RBOC as well as smaller LECs. By building a separate network, the state of Iowa is in effect competing with the LECs in the state. A number of bidders submitted offers to build the ICN, including Iowa Network Services, an AT&T–U S WEST partnership, and MFS Technologies. MFS Technologies was the final winner as the company responsible for the construction of the network. McCleod Telecommunications also won a $28-million maintenance contract. Although the network was built by MFS Technologies, the systems integration arm of MFS, the role of MFS in this project is limited to the construction and maintenance of the network.

The ICN was approved by the state legislature at the end of 1989. At the time, the cost of building the ICN was estimated at $50 million. Since then, network construction incurred major cost overruns. In fact, by 1992, cost estimates had climbed to $300 million. To finance the major increase in cost, the state of Iowa sold about $120 million in bonds.

The high cost of the network was attributed to several reasons:

- The ICN is an all-digital network. Due to the digital nature of the network, the cost per classroom reached $35,000. Had the state of Iowa selected an analog design, the cost per classroom would have been around $20,000.
- While the analog design would have provided the state with 16 channels of video in each local loop district, the digital network provides the local loop district with only three channels on a DS3 circuit.

To use the network, the schools are charged $5 per hour. The remainder of the cost will be subsidized by the state. Each school must also pay $35,000 to $45,000 to install and equip the video classroom.

6.2.1.2   The Future of the ICN

By late 1994, only 10% of the schools (54 schools) in the state of Iowa were connected to the ICN. However, as time goes by, the ICN will be expanded to incorporate a larger number of players. At the next phase, an additional 543 schools and libraries will be connected to the ICN. In addition, there are proposals to open the ICN to hospitals, medical clinics, and federal government offices. Several federal government agencies are considering connecting into the ICN, including the General Services Administration (GSA), the Postal Service, the Department of Veterans Affairs Health and Human Services, the Emergency Management Agency, the Department of Agriculture, the Department of Justice, and the National Academy of Public Administration.

Efforts to expand the ICN are supported by the governor, the Iowa members of Congress, and the media. To expand the ICN network, the state of Iowa has gone out for bids, based on leasing local facilities for seven years. More than 130 proposals have been submitted.

While the creation of the ICN had several shortcomings, other initiatives, such as the NCIH, were more positive. Two major reasons made the NCIH a more positive experience than the ICN:

- In the case of the NCIH, the North Carolina state government acted as the end user. The upgrade of the NCIH was left to the network service providers, which have more expertise in advanced technologies. In contrast, the ICN is managed and upgraded by the state of Iowa.

- In the case of ICN, the state of Iowa competes with the network service providers in the state. In contrast, NCIH does not duplicate the network development efforts of network service providers in the state.

6.2.1.3 Other Statewide Networks

The ICN and NCIH are examples of a statewide network. Other examples of states that have made similar efforts to build statewide networks are Utah, Texas, Georgia, and California [4]. (Chapter 5 described these and many other efforts.)

6.2.2 The Public Policy Role of State Governments

The public policy initiatives of state governments related to the NII and IDL have been shaped by the following drivers:

- Ensuring that universal services, including IDL services, can reach all the citizens of the state;
- Fostering competition among the telephone companies, the cable TV companies, and the AAPs to stimulate the growth of advanced broadband services;
- Enacting legislation to establish funds for fostering educational initiatives.

To accomplish those objectives, state governments have been removing regulatory barriers to competition in the telecommunications industry at the state level and allowing telephone companies to offer discount price to stimulate the growth of IDL.

6.2.2.1 Distance Learning Services at Discounted Rates

State governments have been promoting the growth of IDL networks to reach the disadvantaged students in rural areas as well as in the inner cities through the approval of special pricing arrangements for education. For example, the Mississippi Public Service Commission approved South Central Bell's Distance Learning Service [5]. The service enables accredited schools, colleges, and universities, as well as public libraries to communicate with each other and remote resources via interactive, two-way video and data. The service also enables Mississippi students to access the Internet and other remote educational resources.

Another example of reduced telecommunications rates to educational institutions is provided by the state of Texas [6], as discussed in Chapter 5. The 73rd Legislature enacted H.B. 653 and H.B. 1029, which mandate reduced telecommunications service rates to educational institutions seeking IDL service.

The adopted rule also established the criteria for schools to qualify: A service is used predominantly for IDL when over 50% of the traffic carried, whether video, data, voice, or electronic information, is used for IDL. To comply with the ruling, each local exchange company is expected to file an annual report with the commission indicating the demand for IDL services provided under its IDL tariff.

A third example of a IDL offering at a discounted rate is provided by the state of Maryland [7]. In 1993, the Maryland Public Service Commission approved a discounted tariff on broadband network use for educational purposes.

6.2.2.2   The Role of State Governments in Promoting Telecommunications Competition

The promotion of competition is perhaps the most important role that state governments can play in the short run to create a universal and affordable communications infrastructures. State governments, which have jurisdiction over intrastate communications regulations, have taken major steps to open competition in the local loop. The efforts of the states vary widely from one state to another, but the following are two key steps taken by the more progressive states:

- In the late 1980s, the AAPs were allowed by most states to compete in offering intrastate private lines and dedicated services to IXCs. Among the earlier states to allow competition in dedicated services were New York and New Jersey.
- A few states (e.g., New York, Illinois, Washington, and Oregon), have allowed competition in switched voice services to business customers.

Opening competition has had many positive effects. First, it resulted in the creation of over 40 AAPs. Second, it had a positive effect on the business customers, which have more choices in selecting advanced services, network service providers, and network routes. Third, it had a positive influence on equipment vendors, which are increasing their production capacity to meet the needs of a growing number of service providers. Fourth, competition drove the existing and emerging service providers to expand their broadband networks and to enhance the quality of those networks. The 1996 Communications Act will further increase competition. Competition, however, has a short-term negative effect: the temporary dislocation of employees of the traditional LECs. The short-term negative effect will, however, give way to longer-term positive effects: the creation of the NII and the resulting positive effects for customers and providers alike.

Before such positive results can occur, several regulatory constraints to competition had to be removed. These constraints apply to new entrants as well to existing network service providers.

From the LECs' point of view, they are faced with several regulatory constraints (prior to the legislation):

• LECs were restricted from entering the long distance market. LECs are willing to agree to legislation opening up local competition in exchange for opening the long distance market.
• LECs needed approval for offering new services. They also need regulatory FCC authorization for operation and installation of their networks, while the AAPs do not.
• LECs were subject to pricing, price-cap, or rate-of-return regulation for obtaining right-of-way, while the AAPs are not.
• LECs cannot offer cable TV services (e.g., information ownership) in their individual regions.
• LECs lack pricing flexibility. Any tariffs or changes in tariffs have to be approved by the *public utilities commission* (PUC). AAPs are also required to file tariffs. The tariffs that AAPs file, however, do not have to be as detailed as those offered by LECs.

From the AAPs' point of view, the key impediments to competition as of January 1996 were as follows:

• AAPs lacked cocarrier status in most cases. That means they cannot administer their own NXX numbers, nor are they treated as telephone company peers subject to the same kind of regulations. Instead, they are treated as telephone company competitors, dependent on regulators' special rules. The lack of co-carrier status also means that to interconnect their networks to those of the telephone companies, AAPs are currently subject to collocation rules, which means they have to pay a special tariff for access to the telephone companies.
• AAPs did not have number portability. That means that a business customer of an LEC who wants to buy local services from an AAP instead of an LEC has to get new telephone numbers. That could be a disruption to the customer's business.
• AAPs had not settled with the PUCs in the states where they serve on the charges they should incur for collocation purposes.
• AAPs had to incur the cost of right-of-way to build networks in new cities, about $650/foot.
• AAPs have to pay franchise fees to local governments to receive approval to operate in certain cities. In many cases, the fees are extensive. In one case, MFS went to court requesting a reduction in franchise fees, which the court granted. Legal action, such as that case, are costly to AAPs in terms of time, business lost, and legal resources.

State governments are currently wrestling with all these issues as the new legislation takes hold. They are also looking toward the federal government to play a more active regulatory role.

6.2.3   The Funding Role of State Governments

State and local governments also play a role in funding the *Public Broadcasting System* (PBS), as discussed throughout Chapter 5. This is a topic that has caused heated debate. The question is: Should state and local governments continue to finance public broadcasting? Whatever the outcome of this debate, one thing is certain: Funding for PBS and its affiliated radio stations will continue to decline.

6.2.4   Potential NII and Distance Learning Roles of State Governments

The current roles of state governments as end users, founders, and regulators have been examined. State governments need to reexamine those roles in light of the shrinking resources available to them, the political climate, and their expanding social roles as the federal government continues its decentralization efforts. Consequently, state governments need to reexamine their roles as founders and operators of statewide networks. The ICN example demonstrates the financial burden that a state can incur (and the cost overruns driven) by taking on the responsibility of building and operating such a network. The cost overruns associated with building these networks translate into higher taxes for the citizens of the states where those networks are built.

State governments also need to reconsider their roles as managers of statewide networks for two reasons: (a) by managing these networks, they are competing with the LECs and other network service providers, and (b) state governments are in no position to retain their telecom/datacom experts because the pay scale in the public sector is lower than that in the private sector. Consequently, in the long run, state governments will not be in a position to retain their skilled telecom/datacom managers. Individuals will likely join the telecom/datacom departments of state governments to gain networking and telecom/datacom expertise and then find better paying employment opportunities in the private sector.

State governments that have taken on the responsibility of managing statewide networks can reduce their network management cost by outsourcing the networks to the private sector [11]. Just as many large businesses have outsourced their networks to focus on their core businesses, state governments should consider outsourcing their networks to focus on their core businesses: providing high-quality government services at the lowest possible cost (i.e., taxes).

State governments should also consider the NCIH as a better model of the role of the state governments in creating a statewide network. The state of North Carolina provided a leadership role in the creation of the network while assigning the responsibility of operating and upgrading the network to the more experienced telephone companies.

While the role of the state government in managing statewide networks should be reduced or modified, state governments should continue to play a vigilant role in monitoring the rates that telephone companies charge IDL users. State PUCs should also emulate those states (e.g., Texas) that authorize discounted rates for distance learning services. On the regulatory front, the state PUCs that have not opened competition in the local exchange market should take the necessary steps to do so. The barriers faced by the state government should be removed to accomplish full competition in the next few years.

State governments should also expand their roles as promoters and educators of the NII and IDL. In addition, state governments should consider expanding their roles as end users. Specifically, state governments should accelerate their investments in the creation of information systems that make the state government employees more productive. They should also expand the use of IDL to (re)train welfare recipients in the basic skills they need to join the workforce.

6.3 NII AND DISTANCE LEARNING INITIATIVES OF THE FEDERAL GOVERNMENT

The Clinton-Gore administration has attempted to play an active role in the creation of the NII. The federal government has become a major supporter and funder of telecommunications development. The Department of Education, the Department of Commerce, the Agriculture Department, Health and Human Services, and the National Science Foundation have all increased their spending on IDL in recent years. The goals of the administration in supporting the creation of the NII are as follows:

- To enhance the competitive position of U.S. businesses relative to businesses in other developed countries;
- To foster competition among players in the telecommunications industry;
- To educate those located in rural areas;
- To simulate economic growth.

The signing of the Communication Act of 1996 goes a long way. To accomplish these goals, the federal government has played end-user, regulatory, financing, and educational roles. Each of these roles is examined next.

6.3.1   The End-User Role of the Federal Government

A number of federal government agencies play major roles as end users of information technologies. These include the Department of Defense (DoD), NASA, the Department of Energy, and the NSF. The networks of these departments represent the core components of what has become known as the Internet.

The networks of these departments are being upgraded as part of the federal government initiative called the *National Education and Research Network* (NREN). This initiative is intended to accomplish two purposes: (a) to support the development and implementation of a national broadband network providing researchers and engineers with the bandwidth necessary for applications such as supercomputing, and (b) to test and demonstrate advanced information technologies before they are deployed on a wide commercial scale.

Through the NREN initiative, federal agencies are participating in test-beds located throughout the United States. The following are the key federal testbeds:

- *AURORA*. This testbed network covers the Northeast. Participants include Bell Atlantic, Bellcore, IBM, MCI, MIT, NYNEX, and the University of Pennsylvania. The AURORA testbed is intended to test high-speed multimedia applications [8].
- *BLANCA*. This testbed network is a national testbed. Participants include AT&T, CU at Berkeley, Berkeley Lab, the National Center for Supercomputing Applications, the University of Illinois, and the University of Wisconsin. The BLANCA testbed is intended to test distributed supercomputer applications, new WAN protocols, and ATM switches.
- *CASA*. This testbed network covers southwestern states. Participants in the testbed include CIT, Jet Propulsion Lab, Los Alamos National Lab, MCI, San Diego Supercomputing Center, and U S WEST. The CASA testbed is intended to test supercomputer-access processing power.
- *MAGIC.* This testbed network covers Minnesota and South Dakota. Participants include DEC, L. Berkeley Lab, the Minnesota Supercomputing Center, MITRE, Nortel, Split Rock, Sprint, SRI International, SBC, the U.S. Army Future Battle Lab, the U.S. Army High Performance Computing Research Center, U.S. Geological Survey, and the University of Kansas. The MAGIC testbed is intended to test ATM products and future battle laboratory applications, including interactive realtime visualization.
- *NECTAR*. This testbed network covers Pittsburgh. Participants include Bell Atlantic, Bellcore, Carnegie-Mellon University, and the Pittsburgh Supercomputing Center. The NECTAR testbed is intended to test HIPPI in the LAN, ATM over SONET, and LAN-WAN interconnectivity issues.
- *VISTAnet*. This testbed network covers North Carolina and is one of the drivers behind the establishment of the NCIH. Participants include Bell-

South, GTE, North Carolina State University, and the University of North Carolina. The main applications of the VISTAnet testbed are telemedicine applications.

These testbeds are coordinated by the Corporation for National Research Initiatives, with the exception of MAGIC. The cost of these testbeds to the United States was estimated to be $15 million for a three-year period.

6.3.2 The Funding Role of the Federal Government

Historically, the federal government has played an active role in financing activities to stimulate the creation of the NII and to encourage the growth of IDL. One of these activities is financing PBS. PBS provides educational services, particularly to those that seem to have the most critical needs and that are not served by other service providers. This traditional role is being challenged by the growing budget deficit and by the recent political realignment in the Senate and the House of Representatives. The most likely result of these new financial and political realities is that funding for PBS will continue to decline in the next coming years.

6.3.3 The Regulatory Role of the Federal Government

The primary federal agency with regulatory responsibility is the *Federal Communications Commission* (FCC). One of the roles of the FCC is to ensure a minimum level of quality in the delivery of telecommunications services by imposing federal standards on network service providers. Local franchising authorities, however, have the responsibility for enforcing the federal standards. For example, to ensure affordable universal service, cable programming rates are subjected to federal review and regulation. The federal government establishes a benchmark that must be followed by the by the local authorities. Cable deregulation, however, was one of the goals of the new Communications Act of 1996.

The federal government also fosters the ubiquity of educational programming in general and IDL in particular, by imposing new "must-carry" requirements on the cable TV companies. According to those standards, every cable system must carry one *noncommercial education* (NCE) station. They must also provide space on their systems for local broadcast stations that elect must-carry status.

Through the FCC, the federal government has also played a positive role in fostering competition among the RBOCs and the AAPs at the interstate communication level. Competition initially was limited to special access services but was then extended to include switched access. Several regulatory barriers, however, still face AAPs, the IXCs, and the LECs. These barriers cannot be removed on a nationwide basis without legislative action by the Congress. This action is likely to be forthcoming in the 1995–1996 timeframe. Absent of these legislative initiatives, state governments will continue to take unilateral actions to foster competition. However, this state-by-state approach results in a patch-work of telecommunications policies that may be different from one state to another [10]. Furthermore, the states are limited in what they can do to foster competition. For example, the states cannot lift the line of business restrictions, which prevent the RBOCs from offering long distance and entertainment services.

6.3.4 The Role of Government as Educator

In addition to the administration, a number of federal government agencies and committees are participating in the NII effort, including the following:

- The White House Office of Science and Technology Policy;
- The National Telecommunications and Information Administration;
- The Information Infrastructure Task Force;
- The U.S. Advisory Council on the NII;
- The Subcommittee on Telecommunications and Finance, Energy and Commerce Committee, House of Representatives;
- The Subcommittee on Communications, Science, and Transportation Committee, U.S. Senate;
- The FCC;
- The NSF;
- CNRI.

6.3.5 The Potential Role of the Federal Government

The federal government will be under increasing financial pressures to balance its budget. Consequently, the federal government needs to reexamine its role as a founder of the NII to ensure that the cost-benefit parameters are in favor of the benefits. The federal government should also accelerate the pace of regulatory reform. The Communications Act of 1996 affords regulatory reform that will be taking place in the 1996–1998 time frame.

As an end user, the federal government should accelerate its investments in the creation of information systems that make federal government employees more effective. The federal government should also speed the process of welfare reform, which can be accomplished by, for example, expanding the applications of IDL to encompass the (re)education of welfare recipients. The role of the federal government as educator, facilitator, and promoter should also be expanded.

The 1996 telecommunications bill [9], which has been hotly debated by Congress for the past few years, was passed in February 1996. Despite the high level of contention surrounding this legislation, particularly during the past two years, it eventually passed with little opposition (414 to 16 in the House, and 91 to 5 in the Senate). The bill was signed by the President. To summarize, the bill will provide for the following changes:

- Local and long-distance telephone and cable companies will be allowed to compete in one another's businesses.
- Cable rates will be deregulated.
- Media companies will be allowed to expand their holdings.
- TV station networks, which are currently allowed to reach a maximum of 25 percent of national viewers, are permitted to expand their viewership to 35 percent of the population. Broadcast companies are allowed to own five to eight radio stations in the same market, compared to the current limit of four stations in a large market, and three in a small one.
- "Pornography" will be restricted on computer networks and television. The V-chip set, which will not be available before 1998, will enable parents to censor objectionable shows from their television screens.

The bill does maintain some of the current limitations, however. These include:

- Prohibiting ownership of more than one television station in the same market;
- Prevention of owning a newspaper and either a TV station, a radio station, or a cable outlet in the same market.

Near-term consequences of this legislation will be heightened competition for long-distance and local telephone services, along with mergers among major network operators. The major local exchange companies have been looking at interexchange services for some time. Unlike interactive video services, interexchange services represent a proven revenue source within their domain of expertise (voice services). On the other hand, interexchange carriers want to bypass local exchange companies and connect directly to end users, thereby avoiding the expensive access fees they pay local carriers to interface their networks. (Access fees represent over 20 percent of an RBOC's annual operating revenues). Large mergers are expected as operators move to increase operations integration both horizontally to expand the reach of existing services and vertically to expand into new services. Bell Atlantic and NYNEX were reportedly engaged in serious merger discussions. Other potential partners, because of the proximity of their service territories, include BellSouth and Ameritech, SBC and U S WEST, and Pacific Telesis and GTE. Telephone company acquisitio of or mergers with cable TV MSOs may also be likely. The passage of the telecommunications bill reflects a global trend in the deregulation of public telephone services. In a directive approved by the Euro pean Parliament, European governments were directed to fulfill their pledge t open all telecommunications markets, including voice telephone service, t competition by 1998. Also, alternative networks, such as utility or railroad net works, will be permitted to begin carrying some telecommunications services.

References

[1] "A National Information Infrastructure (American Style)—An Approach to Government Ir volvement in the Development of Advanced ICT," Global Growth Strategies White Pape Apr. 30, 1994, p. v.

[2] Betts, M., "States Bypassing Feds on Information Highways," *Computerworld*, June 20, 199 p. 8.

[3] "Mississippi PSC Approves South Central Bell's Distance Learning Transport Services," *Th Circuit,* May 1994, p. 5.

[4] "Reduced Distance Learning Rates Available for Schools in Texas," NARUC No. 50-199. Dec. 13, 1993, pp. 25–26.

[5] "C&P Telephone and Maryland Announce Distance Learning Project," *The Heller Repor* Vol. IV., No. 11, Sept. 8, 1993.

[6] Minoli, D., *Analyzing Outsourcing, Reengineering Information and Communication System* New York: McGraw-Hill, 1995.

[7] "A National Information Infrastructure (American Style)—An Approach to Government Ir volvement in the Development of Advanced ICT," Global Growth Strategies White Pape Apr. 30, 1994, p. 11.

[8] Tauke, T. J., "1994 in Review: A Regulatory Perspective," *Telecommunications*, Jan. 199 p. 49.

[9] Hankim, Ryan, Weekly Newsletter on Communications Technologies and Services, Februar 2, 1996.

Bibliography

"GA Creates Large Funding Pool for DL Instructional Technology," *The Heller Report*, Vol. IV No.10, pp. 1, 7.

Fugel, J. A., "More Than Meets the Eye: The Iowa Communications Network: A Bad Deal for Every one," *Rural Communications*, Jan./Feb. 1995, p. 18.

"Government or Business: The Developing Battle Over Who Takes the Lead in Distance Learning. *Broadcasting*, Nov. 30, 1992, pp. 42–47.

NII and Distance Learning Initiatives of the Telephone Companies

7

Telephone companies have been active in the creation of broadband network infrastructures and in developing, among other applications, IDL solutions to ride on these infrastructures. This chapter explores those efforts. The two groups of telephone companies that are examined are the LECs, particularly the RBOCs, and the IXCs. For each group, the major players are identified, the key elements of their network infrastructures are described, their IDL initiatives are examined, and their networking services are analyzed. The chapter concludes by discussing the strengths and limitations of the initiatives.

7.1 THE LOCAL EXCHANGE CARRIERS

The efforts of the telephone companies to build the NII and to deliver IDL solutions have been shaped by the following challenges:

- Increasingly intense competition from cable TV companies and the AAPs;
- Slow growth in traditional telephone company sources of revenues (3%–4%);
- Increasing need for new sources of revenues that can offset the revenues lost to competition.

The telephone companies view the creation of broadband network infrastructures and IDL solutions as tools to face these challenges and to differentiate themselves from existing and emerging competitors. To face those challenges, the LECs are taking these steps, among others:

- They are in the process of upgrading their network infrastructures to support broadband applications.

- They have built broadband testbeds to stimulate the demand for broadband services, including IDL.
- They have initiated a number of IDL projects (e.g., Pacific Bell's CALREN).
- They have expanded the range of networking services aimed at residential, corporate, and educational markets.

7.1.1 The Evolving Network Infrastructure of the LECs

The LECs are in the process of building networks infrastructures based on the following elements:

- A fiber SONET/SDH backbone at the physical layer level;
- An ATM platform at the multiplexing (data link) layer level;
- Multiple services at the application layer, including Internet access and multipoint videoconferencing services.

Currently, the geographic scope of these infrastructures is mostly focused on intraLATA communication, although new regulation allows them to enter interLATA markets.

7.1.2 Example of a Broadband Network Infrastructure

An example of a broadband network infrastructure,as discussed in previous chapters, is the NCIH. The NCIH was created at the request of the North Carolina state government, which recognized the potential of broadband technologies in serving the goals of the state to become more competitive, to attract jobs, and to educate its citizens. The state of North Carolina requested BellSouth, Carolina Telephone, and GTE to propose a plan for the construction of a statewide information infrastructure based on broadband technologies [1]. The companies developed proposals that were approved by the state. Figure 7.1 shows the initial network infrastructure built by those LECs in 1994.

At the data link (multiplexing) layer, ATM was the selected technology. At the customer premises, existing interfaces such as DS1 and DS3 are converted to ATM cells by an ATM service multiplexer. The customer-owned ATM service multiplexer has two interface cards. The first card is used for DS3 circuit emulation using AAL Type 1 protocols (see Chapter 3 for a discussion of AALs). This interface card supports traditional digital video communications (newer approaches to the transport of digital video over ATM use AAL five protocols). The second card supports SMDS through AAL Type 3/4 and is used for connectionless data communications. The transmission of ATM cells from customer premises to the ATM switch at the central office is accomplished through SONET facilities at OC-3 speeds.

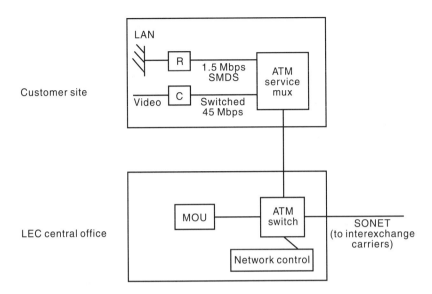

MCU: Multipoint control unit
R: Router
C: Video codec

Figure 7.1 LATA architecture in first phase of the NCIH (1994).

Once the ATM cells reach the ATM switches at the LEC central office, they may take several routes. The cells carrying SMDS traffic are carried to an SMDS handler for connectionless transport via an overlay SMDS network. The cells carrying DS3 digital video signal are converted to analog video signals at the MCU to enable functions such as split-screen video bridging. Interconnection between ATM switches in various LATAs is provided through interLATA SONET facilities provided by the IXCs. When the ATM cells reach their final destinations, the original interface is generated by another ATM service multiplexer.

As of early 1995, 45 customer sites were connected to the NCIH. These sites were served from 12 switches and 7 remote concentrators (see Figure 7.2). The NCIH architecture was expected to evolve by 1996 to that shown in Figure 7.3. The following are some of the expected changes:

- Additional ATM switches will be needed in each LATA.
- Additional remote concentrators will be needed in each LATA.
- MPEG-2 video codecs operating at about 6 Mbps were expected to become available providing high-quality video at one-seventh the original DS3 bandwidth.

- ATM SVC capability based on the ATM Forum UNI 4.0 specifications were expected to be become available.

7.1.3 Distance Learning Initiatives of the LECs

In addition to the NCIH effort, BellSouth and the other LECs have been active service providers in the IDL market. Examples of IDL initiatives abound, as discussed in Chapter 5.

Bell Atlantic and the Maryland state government started moving forward with a plan to bring two-way audio/video IDL to all state public high schools, community colleges, and four-year state colleges [2]. In mid-1993, the Maryland Public Service Commission approved a discounted tariff on broadband network use for educational purposes.

BellSouth has introduced IDL services at discounted rates in several states. For example, in Mississippi, South Central Bell introduced such a service [3]. This service will enable schools, colleges, universities, and public libraries to communicate with each other and with remote educational resources via interactive two-way video and data. The services will also enable Mississippi students to access the Internet.

In California, Pacific Bell announced a $100-million investment in 1994 to help California schools and libraries connect to the NII [4]. This investment is aimed at offering linkages for computer communications and videoconferencing to 7,400 public K–12 schools, public libraries, and community colleges in

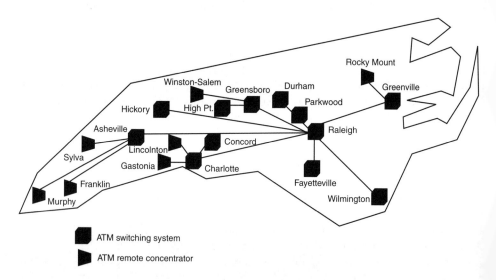

Figure 7.2 Locations of ATM switches and concentrators in 1994.

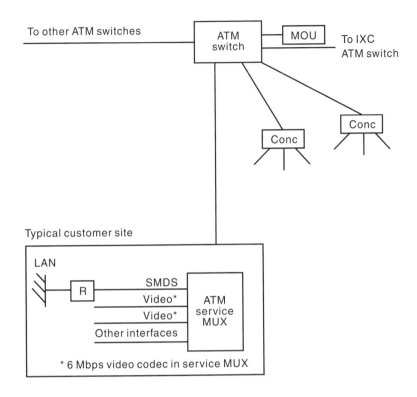

Figure 7.3 The emerging LATA architecture of the NCIH.

Pacific Bell's service territory by the end of 1996. Pacific Bell also accepted a challenge from state lawmakers to lead businesses, regulators, and legislators to an initiative to ensure that every classroom in California is wired for broadband services by the year 2000. Pacific Bell is reported to be willing to wire targeted institutions, install services free of charge, and waive the usage charges for one year after installation.

In 1993, NYNEX and the New York City Department of Telecommunications and Energy announced the introduction of Classnet [5]. This is the first interactive video network created to serve New York City students. Classnet employs two-way video and audio and enables students and teachers in four separate locations to hold classes.

SBC has built IDL networks in all the states it serves, including Arkansas, Kansas, Oklahoma, and Texas. For example, in Kansas, SBC helped nine different schools by building the A-Plus network. This network connected the nine districts through interactive video and audio networks. The network helped those schools expand their curriculum and create flexible scheduling.

U S WEST initiated multiple IDL projects. For example, in Minnesota, U S WEST replaced a microwave system in the Minnesota River Valley with an analog fiber optic network that connects five schools and a technical college. In Colorado, U S WEST assisted the St. Vrain Valley School district in establishing communications links between Lyons and Logmont.

Ameritech has implemented numerous IDL projects. For example, in Ohio it announced a new $1.6-billion infrastructure program that calls for building interactive video IDL networks interconnecting all of the state's 559 high schools, vocational schools, colleges, and universities [6].

7.1.4   Network Services of the LECs

To support the needs of the IDL community at the networking level, the LECs have been expanding the range of network services they offer. These services can be categorized into two groups:

- Services that support the local communities, including K–12 schools, and residences;
- Services that support the business community and large customers, including major universities, large business customers, and government agencies.

7.1.4.1   LEC Services That Support Local Communities

ISDN

One of the major services that the LECs offer to support local communities is ISDN. All the RBOCs have been actively deploying ISDN. During the next five years one can expect ISDN to be, finally, widely deployed. The next few paragraphs give a sense of the deployment (plans) at press time.

Ameritech has been actively deploying ISDN [7]. About 75% of Ameritech's lines were planned to have the capability to be activated as ISDN by 1995. Ameritech also was planning to offer multirate ISDN, a variation on the ISDN PRI, which can operate at speeds in multiples of 64 Kbps. Bell Atlantic also supports the IDL industry by offering ISDN. In 1993, Bell Atlantic started offering individual line ISDN to small businesses in Virginia, West Virginia, Maryland, and Washington, D.C. ISDN BRI was expected to become regionwide and available to residences.

BellSouth deployed ISDN extensively to serve the needs of a number of markets, including the IDL market. According to industry forecasts, almost

60% of BellSouth lines were planned to have the capability to be activated as ISDN by 1995. BellSouth tariffed two versions of ISDN throughout its nine-state territory: ESSX ISDN (Basic Rate/Centrex) and MegaLink ISDN (PRI). BellSouth also trialed an individual line ISDN [9]. In addition, BellSouth introduced a tariff allowing a business line to homes at residential prices.

Pacific Bell has been deploying ISDN to serve the needs of residential as well as IDL markets. In addition to ISDN, Pacific Bell offers several new services to support the needs of the IDL industry, including *Switched Digital Service 56* (SDS 56), a circuit switched digital service.

NYNEX has been expanding its ISDN capabilities by deploying ISDN-capable central office switches in an increasing number of central offices throughout New York. NYNEX also offers ISDN PRI. SBC has announced plans to rapidly expand the availability of ISDN. U S WEST is also increasingly offering ISDN throughout the Northwest.

Video Dial Tone

Recognizing the need of local communities for bandwidth that exceeds that delivered by ISDN, the RBOCs have been trialing VDT technologies [18]. These technologies and services should become available and acquire importance during the next five years. VDT is an emerging information exchange and exchange access service that is being offered to support VIUs, including IDL receivers and providers, in accessing video and multimedia applications offered by VIPs. Each VIU subscribing to a VDT service is able to establish either a single session or multiple sessions depending on implementation. Initially, VDT services offered by the LECs will support one-way video under user control. Video distribution may be point-to-point or point-to-multipoint, depending on the application. The implementation of VDT networks can be based on ATM/SONET (ATM supports the switching function, while SONET provides the transmission function) or other hybrid technologies, such as HFC or (perhaps less likely) ADSL. (See Chapter 1 for a description of ADSL and HFC.)

VDT may find applications in institutional delivery of video, for example, to schools. Cable TV companies already have a high market share: 78% of all public elementary and secondary school districts and 62% of all public school locations use cable for distance learning applications (in addition, almost half the schools have a satellite dish for increased service capabilities). However, the cable TV companies have as a goal to add the capability to handle two-way switched communications; VDT provides such capability. Table 7.1 depicts statistics on the number of schools in the United States [18]. Studies show an expected classroom penetration for interactive video of 1.5% in 1995, 10% in 1998, and 25% in 2003.

Table 7.1
Distribution of Schools by RBOC

Public Schools	Elementary	Secondary	Postsecondary
Total schools	62,350	20,375	10,875
Ameritech	18.9%	18.7%	20.8%
Bell Atlantic	12.3%	11.4%	23.3%
BellSouth	17.4%	15.6%	16.5%
NYNEX	11.5%	5.9%	14.7%
Pacific Bell	8.9%	10.0%	16.9%
SWBT	14.2%	17.9%	3.6%
U S WEST	16.8%	20.5%	04.2%

Private Schools	Elementary	Secondary	Postsecondary
Total schools	16,475	2,100	1,775
Ameritech	22.9%	19.9%	17.3%
Bell Atlantic	16.6%	17.8%	14.7%
BellSouth	15.8%	12.1%	18.3%
NYNEX	12.5%	19.9%	20.2%
Pacific Bell	11.9%	12.2%	7.8%
SWBT	8.0%	7.6%	10.2%
U S WEST	12.2%	10.7%	11.5%

In 1994, Ameritech announced that it was planning to spend $4.4 billion to offer video services to 6 million customers by 1999. The company filed applications with the FCC to deliver VDT services to 1.26 million customers in its region in 1994, including residential customers, libraries, schools, and hospitals. Ameritech plans to add 1 million customers per year to reach the targeted 6 million total by 1999. Customers will be offered a number of services, including video on demand, interactive education, games, entertainment, and information services. VDT services offered by Ameritech are based on the HFC technology [19].

Bell Atlantic is reportedly planning to spend $15 billion through 2002 on replacing its twisted-pair access plant (local loop) with an optical fiber plant. Bell Atlantic also plans to reach 1.2 million households in 1995 and 9 million households in 2000 in the top 25 markets. These households will be offered a number of VDT services. At the first stage, these VDT services will be based on the ADSL technology. Bell Atlantic has been conducting an ADSL-based trial in Virginia since 1994; the goal of the trial is to offer twisted-pair-based VDT services to over 300,000 homes by 2003. In addition to its ADSL-based activities, Bell Atlantic has built a two-way FTTC system in Morris County, New Jersey. Bell Atlantic's New Jersey trial initially employed HFC equipment from Broadband Technologies, Inc. [19].

NYNEX has been conducting a number of VDT trials. In early 1994, NYNEX announced that it will deploy a broadband HFC network reaching 1.5 million to 2 million consumers by the end of 1996. The first 60,000 lines were planned for Warwick, Rhode Island. Initially, NYNEX will deploy the broadband network as an overlay to the existing public switched network. This means that the broadband network will carry video while the public switched telephone network will continue to support traditional voice services. After 1996, NYNEX is publicly committed to deploy the HFC network at the rate of 1 million lines per year. NYNEX has not, however, committed to universal broadband service. This is because NYNEX does not find available broadband technologies economic enough to justify deployment in rural areas. NYNEX is currently considering evaluating a number of broadband technologies, such as wireless and ADSL [19].

In the mid-1990s Pacific Bell announced a $16-billion plan aimed at providing VDT along with other advanced services to about 5 million homes by the year 2000. As an initial step toward that goal, Pacific Bell requested permission from the FCC to provide video services over the telephone network to 1.3 million homes in California. Initially, Pacific Bell will rely on the HFC technology. The company indicated that it was going to purchase HFC equipment from AT&T and video servers from Hewlett-Packard. Initial targets of VDT services are Los Angeles, San Francisco, San Diego, and Orange County. In early 1995, Pacific Telesis announced an early VDT trial in Malpitas, California, serving 1,000 customers by the end of 1994. The trial allows customers to view video in multiple windows on their TV screens and to preview programming [19].

In early 1994, SBC announced a video service trial in Richardson, Texas. The trial, which was expected to last one year, is based on an HFC architecture that supports video and data services to 100 to 400 nodes per home. These services include pay per view and video on demand. The company is also investigating interactive games, shopping, and education. The architecture was expected to rely on ATM switches to support services on demand. The video servers were to be placed at the central office as well as in vaults within the neighborhoods [19].

U S WEST has been conducting VDT trials in its own territory. For example, U S WEST was planning a VDT trial in Omaha, involving 60,000 subscribers. The trial is based on the HFC technology. Beyond the trials, U S WEST has plans to offer VDT services in at least 15 areas. U S WEST is reported to have committed $2.5 billion to the creation of VDT services [19].

7.1.4.2 LEC Network Services to Large Business Customers

In addition to VDT and ISDN, the LECs have been increasingly deploying broadband data services to serve the needs of business customers, including distance learning customers. The following are some of their key activities related to broadband services as of press time (more services are expected in the future):

- Bell Atlantic offers SMDS and FRS in addition to an FDDI-based version of NMLIS.
- BellSouth trialed NMLIS in mid-1993. BellSouth planned to transition NMLIS to backbones based on ATM switching [10]. BellSouth also offers SMDS and FRS as interim high-speed data networking platforms. Bell-South's long-term plans were to migrate these two services to an ATM platform.
- In 1993, NYNEX introduced NYNEX Enterprise Services [11], a group of private lines supported by bandwidth on demand and network management capabilities. NYNEX also introduced FRS using Nortel's DataSPAN platform. In addition, NYNEX offers fractional DS1 services at 128, 256, 384, 512, and 768 Kbps [12].
- SBC offers SMDS, FRS, and circuit switched videoconferencing services at 56 Kbps, 384 Kbps, and 1.5 Mbps [13].
- U S WEST has deployed SMDS and FRS and has introduced a transparent LAN service (TLS) [14].
- Ameritech deployed ATM switches in Chicago and Milwaukee as early as 1993. Ameritech's initial offerings on its ATM platform emphasize SMDS and include a range of bandwidths from 64 Kbps to 384 Kbps [3]. In 1993 Ameritech also introduced the Ameritech LAN Interconnect Service (ALIS). In addition, Ameritech introduced a 384-Kbps fractional DS1 service throughout its five-state region.
- BellSouth has also introduced fast packet services, such as ATM-based CRS and SMDS.

The second step that telephone companies are undertaking in creating the NII is to stimulate the demand for NII-based broadband services through the creation of broadband testbeds. An example of such testbeds is CALREN, a project launched by Pacific Bell in 1993. The goal of Pacific Bell associated with CALREN is to promote the development of applications that improve education and healthcare delivery, enhance government processes, stimulate business development and job creation, and strengthen the competitive position of the state of California.

Initial CALREN participants in the testbed include a number of major R&D firms, computer companies, and universities, as shown in the following list:

- Apple Computer, Cupertino;
- DEC Systems Research Center, Palo Alto;
- HP Laboratories, Palo Alto;
- Lawrence Berkeley Laboratory, Berkeley;
- Lawrence Livermore National Lab, Livermore;
- NASA Ames Research Center, Moffet Field;
- Sandia National Laboratories, Livermore;
- Silicon Graphics, Mountain View;
- SRI International, Menlo Park;
- Stanford University, Stanford;
- Sun Microsystems, Computer Corporation, Mountain View;
- University of California, Berkeley;
- Xerox, Palo Alto, Research Center, Palo Alto.

7.1.5 Strengths and Limitations of the LEC Initiatives

As discussed in the preceding section, the LECs are building a broadband network infrastructure and are offering services. The strengths of the LECs in this arena are:

- *A high-performance backbone.* The network infrastructure that the LECs are building supports broadband connectivity, which is key to video-based IDL.
- *An expanding range of applications.* The LECs are offering an expanding range services to support videoconferencing and data conferencing solutions.
- *An expanding ISDN network.* The LECs are increasingly making ISDN services available to local communities and to business customers.
- *Affordability.* The network services that the LECs offer are becoming increasingly affordable from a financial perspective.

These factors make the LEC networks an important component of the NII. The LEC networks, however, cannot be considered as a complete substitute to the NII for the following reasons.

- *Lack of support for rural communities.* The major LECs have been under-investing technologically in rural communities because they do not consider the deployment of advanced networks in those communities to be profitable.

- *Limited broadband service support for K–12 school.* While the LECs, specifically the RBOCs, are experimenting with VDT and expanding their broadband network infrastructures, these broadband services are largely targeted at residences. Other services are aimed at universities, large business customers, and government agencies. K-12 schools do not yet receive the full benefits of broadband networking capabilities.
- *Lack of universal service to K–12 schools.* LEC services are not universally available. To overcome that problem, the United States Telephone Association (USTA) proposed that the FCC let the LECs contribute 1% of their annual interstate revenues to an education fund for three years starting from 1995. Under the 1% plan, a small telephone company would contribute about $320,000 a year, a large RBOC would contribute $32 million [15]. The voluntary fund proposed by the USTA could be used by schools to connect to the NII, without the requirement to use a specific technology or service provider.
- *Lack of advanced Internet services.* While some RBOCs are beginning to deliver Internet access services, these services are not yet at an advanced stage of development.
- *Lack of advanced integrated capabilities.* The RBOCs do not yet offer an integrated set of service capabilities encompassing the upper communications layers (i.e., application-level functionality).

7.2 THE INTEREXCHANGE CARRIERS

The IXC market continues to be dominated by the three major players: AT&T, with approximately 63% of the long distance market; MCI, with about 20%; and Sprint, with about 17%. While the positions of the largest three players are relatively stable, the market positions of the smaller players are constantly changing. This is attributed to a wave of mergers and acquisitions that is sweeping the market.

7.2.1 Challenges Facing the IXCs

The efforts of the IXCs to build the NII are shaped by the following challenges:

- Increasing competition from some AAPs, although in some cases there is a joining of forces (e.g., Sprint and TCG);

- Slow growth in traditional telephone company sources of revenues (3%–4%);
- High access cost, which accounts for roughly 45% of their revenues. (Access cost is the money that they have to pay to the LECs to complete the calls they carry accross LATA boundaries.)

7.2.2 The Network Infrastructure of the IXCs

To be in a better position to serve the needs of their customers, including those in the IDL industry, the IXCs have been building a broadband infrastructure. The three key elements of the infrastructure at the networking level are: (a) fiber-optic technology, (b) standardization of the transport via SONET/SDH, and (c) ATM.

The structural changes affecting the interexchange industry are not limited to the design of their networks but also encompass the scope of those networks. The scope of some of the IXCs' networks is expanding from the interLATA level to the intraLATA level. Two examples of the expanding scope of the IXCs are the networks of MCI and Sprint. The two IXCs are aiming for end-to-end solutions; however, their strategies in reaching those goals are different. MCI is in the process of expanding its network scope beyond the interexchange network to encompass the local market through the creation of MCI Metro, a recently created AAP subsidiary of MCI. Sprint, on the other hand, has reshaped its network scope when, in 1995, it officially entered into an alliance with TCG and with three cable TV companies, including TCI, Comcast, and Continental. This alliance will have the following benefits to Sprint, its allies, and customers:

- Reducing access cost to business customers in over 45 cities through TCG access services;
- Offering business customers one-stop shopping for intraLATA and interLATA services;
- Enabling Sprint to access the marketing skills of cable partners;
- Enabling Sprint to access tens of millions of cable TV customers (representing one-third of cable TV subscribers);
- Opening new distribution channels for multimedia products and services;
- Providing small business customers with a single point of contact;
- Providing global customers with a one-stop partner;
- Relying on broadband network of cable TV partners;
- Selling telecommunications products under the Sprint brand name;
- Packaging Sprint's long distance service with cable companies' entertainment services.

7.2.3 Services Offered by the IXCs

AT&T, MCI, and Sprint have several similar product characteristics: (a) they support fastpacket services and circuit as well as switched services, and (b) they are increasingly offering application-level solutions.

7.2.3.1   Network Services Offered by the IXCs

The IXCs offer a range of circuit switched services and fast packet services. This information is time dependent, but it offers a snapshot view of what was available at press time. For example, AT&T offers a number of circuit switched services as part of the Accunet family, including switched 56-Kbps, switched 64-Kbps, contiguous switched 384-Kbps, and switched DS1 services. Distance learning receivers and providers can access these services through PRI ISDN. AT&T also offers FRS as part of its Interspan family of products. The access speeds of FRS are up to 1.5 Mbps. AT&T also developed X.25-FRS interoperability under the Interspan umbrella.

MCI offers a wide range of network services to support the needs of the residential market, including switched 56 Kbps and switched 64 Kbps. Both require ISDN access for switched data services. MCI also offers switched DS1 services and switched DS3 services. To support the networking needs of the business market, MCI plans to deploy an ATM-based platform to support FRS, SMDS, and CRS. At press time, MCI supported SMDS and FRS.

Sprint offers switched 56 Kbps and switched 64 Kbps. Sprint did not have a switched DS1 offering at press time; it did have an $n \times 56/64$-Kbps service. To support the needs of business customers, Sprint was the earliest adopter of SONET technologies among the IXCs. Sprint was also among the early adopters of frame relay technology. Sprint introduced the service in 1991 and made it available at all domestic *points of presence* (POPs), the United Kingdom, and Japan. Sprint offers FRS over a frame relay network and CRS over an ATM network.

7.2.3.2   Application-Level Services of the IXCs

The IXCs have been expanding their offerings from the networking and transport levels to the applications level. Through application solutions, the IXCs are in a better position to serve the needs of users for the following distance learning–related needs: videoconferencing services (both one-way and two-way), access to the Internet, and groupware solutions (such as the public-based Lotus Notes solutions offered by AT&T). Some examples follow of the application-level solutions offered by the major IXCs at press time.

Table 7.2 summarizes key AT&T products (prior to the "trivestiture") for video and multimedia at the functional level. Specifically, in the mid-1990s

AT&T and Lotus Development Corporation announced a strategic alliance to deliver integrated messaging to integrate telephony, groupware, and electronic mail. AT&T's WorldWorx Solutions is a family of products and services that provide simultaneous, real-time interactive voice, video, and data communications among two or more people. WorldWorx Solutions combine the resources and product offerings of several AT&T business units into a single, comprehensive set of solutions.

Table 7.2
AT&T Video and Multimedia Products

End-user products	Vistium Multimedia System
	Interactive TV (prototype)
	Digital HDTV (Grand Alliance)
	Computer product line with integrated communications
Customer and network infrastructure products	Cell relay/ATM broadband switch
	Fiber-in-the-loop products
	Video server
	Compression equipment family
High-bandwidth transport services	WorldWorx
	ISDN-based digital services
	FRSs
	ATM/CRSs
	NetWare Connect Service
Content hosting and navigation services	AT&T Network Notes
	Information Access Service
	Internet Network Information Services
	PersonaLink Services

In support of the "any time, any place, any kind" goals, including services in support of distance learning, AT&T has started delivering a set of specialized services aimed at mobile communications, game players, and publishers [16]. To accomplish that, recently AT&T started PersonaLink mobile communication services; it acquired ImagiNation Network game-playing service; it acquired Interchange electronic publishing network; and it provides Internet access through Interspan service.

- *AT&T PersonaLink Service.* This service is aimed at users of handheld devices, such as Sony's Magic Link; the ultimate goal is to achieve total transparency between wired and wireless connections. AT&T scrapped its own device, Eo, but the company plans to allow laptop computers to use PersonaLink. PersonaLink Service is an "easy and enjoyable way," according to AT&T, to manage users' daily communication needs using *personal digital assistant* (PDA) devices. Mobile software programs, called "intelligent assistants," help in carrying out their owners' instructions.
- *ImagiNation Network.* This service enables PCs all over the United States to play chess, checkers, golf, poker, aerial dogfights, fantasy role playing, and other games. Users of this network indicate that it is the "chat" session, online events, and electronic mail that they value the most at this time. AT&T already had a 20% stake in The ImagiNation Network prior to the purchase deal with Sierra On-Line, Inc. Interactive entertainment services, such as the ImagiNation Network, let people meet and form relationship through games and other activities they enjoy together over the network.
- *AT&T Network Notes.* In 1994, AT&T and Lotus Development Corp. announced plans to make Lotus Notes available on AT&T's public network. This service makes the widely used Lotus Notes software a tool for workgroup collaboration, is used by 3,200 companies worldwide, and is available to businesses on a distributed but interconnected basis. Through this service, AT&T makes collaboration among businesses easier by introducing networking capabilities to existing business software environments. Through AT&T Network Notes, business partners, suppliers, and customers can collaborate on the preparation of documents that include text, still images, and video.
- *AT&T NetWare Connect Services.* Offered through an alliance of AT&T and Novell, Inc., these new services will use AT&T's and Novell's directory service and internetworking technologies to form a global network of business networks, giving businesses the ability to share applications and communicate with partners and suppliers.
- *InterNIC Directory and Database Services.* This service helps people navigate through the Internet. Through keyword searches, white and yellow page directories, library catalogs, and data archives, InterNIC helps users find resources, institutions, and other Internet users.
- *AT&T WorldWorx Solutions.* This service lets people at multiple locations share applications or data and image files while participating in a call that includes voice and video. WorldWorx Solutions work with Vistium and

other video systems. WorldWorx Solutions is a series of products and services that bring people together in ways not possible before, allowing them to see one another, talk to one another, and share files and data. Designed for business and institutional use, AT&T WoldWorx Solutions are available in the United States and 22 other nations in Asia, Europe, and South America. AT&T is packing applications for WorldWorx in areas such as distance sales and support and collaborative work teams. AT&T has over 20 partners for WorldWorx Solutions, including Apple Computer, Compression Labs, Hewlett-Packard, IBM, Insoft, Intel, Lotus, Novell, PictureTel, SGI, Sun Microsystems, and VTEL.

- *AT&T WorldWorx 800 Service.* This service combines the ease and convenience of toll-free calling with the power and intelligence of voice, data, and image applications by providing toll-free connectivity for ISDN BRI and other digital interfaces.
- *AT&T Vistium Solution.* This combination of hardware and software enables videoconferencing and simultaneous collaborative file sharing over an ISDN BRI or other digital line using a 386 or 486 computer and telephone equipment. The Vistium product line includes the AT&T Vistium Personal Video System 1200, the AT&T Vistium Personal Video System 1300, and AT&T Vistium Share Software Professional.
- *AT&T Intuity System.* This advanced AT&T Global Business Communications System messaging system for use with PBXs lets users share multimedia messages, add media to a message, forward or leave video messages, and convert electronic messages to voice.

MCI has also introduced a set of application solutions, under the NetworkMCI Business umbrella. This software is an integrated information and communications package. It supports existing CPE, and it is designed to be easy to use, customizable to user needs, and inexpensive ($100 per desktop and a monthly cost of $50). The following are the key elements of the package:

- *InfoMCI* is a personalized information and automated news service that monitors more than 180 news sources and databases to deliver twice-daily on-screen summaries of critical news items.
- *MarketplaceMCI* is a multimedia online catalog and business purchasing service that enables businesses to distribute product information and sell products on line.
- *InternetMCI* is an MCI extension of NetworkMCI that will provide users with access to the Internet through a GUI.
- *FaxMCI* enables businesses to send and receive faxes from a PC in a Windows application. Its features include fax broadcast capability, an address book, and scheduling options.

- *E-mailMCI* relies on the global MCI mail network. It enables users to communicate with LAN-based systems (e.g., cc:mail, Netware MHS, and Microsoft mail) by working as a client on a LAN.
- *ConferenceMCI* is a document-sharing application, enabling business users in different locations to update and review the same document in real time.
- *Desktop videoconferencing* is an optional feature of the package that enables two users to see each other while reviewing a document. Communication takes place over either regular phone lines or ISDN lines. This application is compatible with MCI's multipoint videoconferencing service, VideoNet.
- MCI also offers a turnkey solution to VideoNet customers, including newsite installation, CPE maintenance and support, online troubleshooting, and real-time remote diagnostics.

Figure 7.4 illustrates the key features that MCI VideoNet supports.

MCI can support user needs for VideoNet services at speeds ranging from 112/128 Kbps to DS3. At the low-speed end of the bandwidth range, MCI VideoNet services are supported by two types of digital services: switched 56 services and ISDN.

Switched 56 Services

MCI most commonly uses two digital circuits providing 112/128 Kbps to support user needs at the low end of the speed range. While the picture quality of

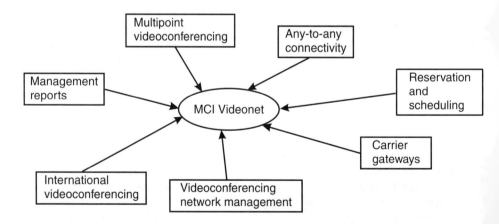

Figure 7.4 Videonet features.

the 112/128-Kbps video transmission is not as clear as the resolution provided by broadcast TV, it is sufficient for some videoconferencing applications, such as informal meetings between small groups of people and lower-end distange learning (e.g., corporate or postsecondary). To establish a videoconferencing connection with other locations, an MCI corporate participant dials the number of the other party (call setup is based on the North American dialing plan). When the circuit switch receives the call, a circuit is established for the duration of the call. This circuit is torn down as soon as the call is completed (both the originator and the receiver, however, need a special digital loop).

Switched 56/64 Services offered by MCI are supported by two platforms: Vnet and MCI Vision.

- *Vnet* is MCI's virtual private network service, which provides large and multilocation customers the benefits of a private network without the expense and administrative burdens of ownership. Vnet switched 56/64-Kbps service provides large customers with the same usage-sensitive rates and discounts as those associated with Vnet voice calls. With Vnet switched 56/64, the same virtual network that supports large customers' voice needs can also support videoconferencing as well as data applications. Vnet's customizable features are available under switched 56/64 service, including on-net and off-net numbering plan and dialing. All Vnet volume discounts and pricing plans also apply to switched 56/64 usage.
- *MCI Vision* is MCI's outbound and card long distance service targeted at small to midsize customers. Vision enables customers to incorporate usage into volume discounts, making it more affordable and convenient to conduct videoconferencing under Vision. As with Vnet, customized features remain available under Vision switched 56.

ISDN

MCI supports ISDN services wherever they are available. At the high-speed end of the bandwidth range, MCI VideoNet Services are supported by two types of digital services: switched services and DS1.

- *Switched services.* MCI combines six individual 56/64-Kbps circuits to support videoconferencing bandwidth needs at the 336/384-Kbps bandwidth level. This frequently used bandwidth level begins to approach full-motion video quality, with movements of participants appearing almost natural. This bandwidth is considered acceptable for most formal or high-level videoconferences.
- *DS1 private lines.* MCI VideoNet can be supported through private DS1 lines. A private DS1 line is a dedicated service that operates at 1.544

Mbps. This service provides a communication link between two locations through the establishment of a physical connection. Two versions of private DS1/DS3 lines are currently offered by network service providers, including MCI: nonchannelized private networks and channelized private networks.

- *Channelized private lines.* Time division multiplexers are currently the most prevalent types of equipment supporting the channelization of private DS1 lines. A time division multiplexer can combine voice, data, and video signals and then send those signals sequentially at fixed intervals. This is accomplished by subdividing a DS1 channel into 24 DS0 channels (64 Kbps); multiple channels can be allocated to one type of traffic (e.g., voice).
- *Nonchannelized private DS1 lines* carry an aggregate 1.544 Mbps in an unslotted manner. In this case, the full DS1 payload is used for the transmission of one type of traffic, which could be data or video. To deploy an unchannelized DS1 solution in support of videoconferencing applications, a codec is used at the locations participating in a video conference.

MCI also supports user needs for VideoNet services at the DS3 bandwidth level. These needs are supported through either private DS3 lines or through switched DS3 services. DS3 speeds provide broadcast-quality videoconferencing for demanding high-performance applications.

At the physical level, MCI supports three types of access options to establish a VideoNet connection:

- *DS1 digital gateway access.* MCI digital access service allows MCI customers to have switched 56 services, private-line data services, and inbound or outbound services on the same DS1 facility.
- *Switched digital access.* This access method allows MCI to access the MCI switched 56 by way of the local carriers switched 56 Kbps offerings.
- *ISDN access types.* These access methods support both MCI switched 56 and MCI switched 64 services. Two ISDN interfaces are available to customers: BRI and PRI.

Sprint also has a set of application solutions. Sprint was the first IXC to offer commercial TCP/IP services and gateways to the Internet through Sprint-Link. In 1994, Sprint expanded its SprintLink Gateway into Global SprintLink. As part of its services, Sprint is responsible for the operation and maintenance of network routing as well as maintenance of the physical network. SprintLink also offers an array of navigational tools and database extensions. In addition, Sprint-Link Plus offers enhancements to Sprint's basic offering, including firewalls and security features.

7.2.4 Strengths and Limitations of the IXCs' Efforts

The IXC infrastructure and services have the following strengths:

- *A high-performance backbone.* The network infrastructures that the IXCs are building support broadband services.
- *Integrated Internet service offerings.* The IXCs have been offering extensive Internet offerings as part of a complete package.
- *Affordability.* The network services that the IXCs offer are becoming increasingly affordable. The affordability of these services is driven by the increasingly competition among the IXCs.

These factors make the IXC networks an important component of the NII. The IXC networks, however, cannot be considered as a complete substitute to the NII for the following reasons:

- *Limited distance learning initiatives.* As recent entrants to the local access market, the IXCs have not been as active in initiating IDL programs to serve the needs of local communities.
- *Lack of support for rural communities.* Like the LECs, the IXCs have not yet extended their broadband capabilities and offerings to rural communities.

7.3 SUMMARY

This chapter indicated that an increasing number of network service providers have initiated IDL projects or have offered an increasing number of services. Those services and initiatives provide IDL users with an increasing range of choices in meeting their needs.

The LECs have been active in building a broadband network infrastructure, and delivering IDL services as well as associated network services. The LECs could further increase their contribution to the IDL market by taking several actions. First, they could reduce the cost associated with rebuilding their infrastructures by creating condominium systems that they share with other network service providers. This is analogous to two separate businesses (tenants) sharing the same building and all the utilities associated it (water, utilities, information systems), while conducting their separate businesses, formulating their separate strategies, and delivering their own set of services. Second, LECs should continue to build broadband and/or multimedia testbeds. This is an example of an effort that provides the potential users with the opportunity to learn about new technologies. It is also an important tool that an

RBOC can rely on in stimulating the demand for new broadband services. The LECs might also want to consider offering an integrated range of service offerings, analogous to those offered by the IXCs.

This chapter also covered the roles of IXCs as providers of IDL solutions. These companies offer the IDL community an expanding range of video services, as well as networking services and integrated service solutions. The chapter also indicated that the traditional boundaries limiting the role of the IXCs to interLATA limits are gradually becoming less rigid. The former structure of the IXCs is being replaced by a new structure, which is still in a fluid state of development. The emerging structure includes the following groups of players:

- *Global alliances.* The best example of a universal alliance is the alliance among Sprint, TCI, and TCG. This alliance will provide the business as well as local community users with one-stop shopping for interLATA and intraLATA services.
- *Global providers.* An example of global providers is MCI. MCI distinguishes itself by offering advanced applications solutions, which fall under the NetworkMCI umbrella. Within five years, MCI is expected to have a ubiquitous network throughout the United States
- *The interLATA providers.* These are the remaining IXCs, which are limited to the older role of the IXCs providing interLATA connectivity. Those companies, however, have been expanding their local access suppliers to encompass the AAPs.

The net impact of these realignments in the telecommunications industry on the IDL industry is that distance learning users will have more communication choices. They will also be able to choose among an expanding range of services.

References

[1] Larry W. Grovenstein, L. W., C. Pittman, J. Simpson, and D. R. Spers, "NCIH Services, Architecture, and Implementation," *IEEE Network*, Vol. 8 No. 6, Nov./Dec. 1994, pp. 18–19.

[2] "C&P Telephone and Maryland Announce Distance Learning Project," *The Heller Report*, Vol. IV., No. 11, Sept. 8, 1993.

[3] "Mississippi PSC approves South Central Bell's Distance Learning Transport Services," *The Circuit*, May 1994, p. 5.

[4] "Students Get Ticket to Ride on Superhigway," *Direct Marketing*, Apr. 1994, p. 10.

[5] "Education's Electronic Superhighways," *Yankeevision Consumer Communications*, Vol. 10, No. 6, May 1993, p. 10.

[6] *Communications Daily*, Apr. 13, 1993, p. 7.

[7] "ISDN Deployment: The Tortoise Makes Its Move," *Data Communications*, Sept. 1993, p. 17.

[8] "Searching for Applications, BellSouth Offers Single-Line ISDN," *Telephone Week*, Jan. 18, 1993, pp. 2–3.

[9] Minoli, D., *Video Dialtone Technology, Approaches, and Services: Digital Video over ADSL, HFC, FTTC, and ATM*, New York: McGraw-Hill, 1995.

[10] Minoli, D., "Video Dialtone Overview," *Datapro Report* 1090, May 1995.

[11] "Transparent LAN Ties," *Communications Week*, Feb. 22, 1993, p. 1.

[12] "NYNEX Fights to Regain Lost Customers With Innovative Service," *Telephone Week*, Apr. 19, 1993, p. 4.

[13] "Bell's Limited Frac T-1 Supports Strands IXCs," *Intelligent Network News*, Dec. 24, 1992, p. 1.

[14] Wallace, B., "Carriers Expand the Reach of Frame Relay Services," *Network World*, Mar. 22, 1993, p. 4.

[15] "U S West Introduces LAN Interconnection Service," *Network World*, Aug. 24, 1992, p. 19.

[16] Gerwig, K., "USTA Supports Distance Learning," *Interactive Age*, Mar. 13, 1995.

[17] Minoli, D., "AT&T Vendor Profile," *DataPro Report on Convergence Technologies*, Apr. 1995.

Bibliography

Minoli, D., "MCI Communications Corp. Videoconferencing Services," *DataPro Report* 6060, May 1995.

The Role of the Cable TV Companies and the AAps

8

The cable TV industry is the main "rival" of the traditional telephone companies in the creation of the NII. After a number of failed attempts to create alliances with telephone carriers, cable TV companies have decided to build elements of the NII without help from the telephone companies. Two major exceptions are the Time Warner/U S WEST alliance and the Sprint/TCI et al. alliance. The goals of the cable TV companies in contributing to the creation of the NII are driven by the challenges they are facing and by emerging market opportunities. This chapter explores the goals of the cable TV companies as well as the strategies they are employing to achieve those goals.

The AAps are another group of competitors to the LECs. AAps have their own set of business goals and their own network deployment strategies and service deployment strategies. This chapter explores the goals and strategies of the major AAps in building their own infrastructures and delivering an expanding range of services that can support the distance learning market. Some of these AAps have established close relationships with some major cable TV companies.

This chapter goes on to explore the resulting restructuring of the telecommunications industry. It concludes by examining the extent to which the cable TV companies and the AAps are contributing to the creation of the NII and the critical factors that are shaping the degree to which they can contribute in the future.

8.1 THE CABLE TV INDUSTRY

The cable TV industry includes over 1,200 multiple systems operators (MSOs). The largest two cable TV providers, TCI and Time Warner, have over 30% of the cable TV market share. Figure 8.1 shows the key cable TV companies and the number of residential customers that each serves.

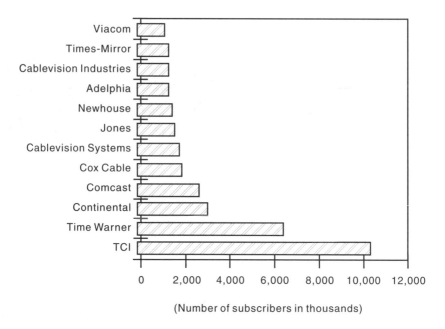

(Number of subscribers in thousands)

Figure 8.1 The top 12 cable TV companies.

8.1.1 Major Challenges Facing the Cable TV Industry

The cable TV industry faces the following major challenges as its participants enter the second half of the 1990s:

- *A fragmented industry.* There are currently over 12,000 cable TV systems in the United States, reaching 60% of U.S. households. Unlike the LECs, which generally have contiguous geographic boundaries in most of their territories, the cable systems of the cable TV companies are dispersed throughout the country.
- *Increased competition.* The cable TV companies are faced with increasing competition from the telephone companies as well as from providers of new technologies, such as direct broadcast satellite (DBS).
- *An unfavorable regulatory environment prior to the new Communications Act.* The cable TV companies faced an unfavorable regulatory environment, which has forced them to roll back their rates.
- *Unreliable and inefficient analog network.* While the traditional analog cable TV networks may be appropriate for the delivery of entertainment services, the analog networks lack the key features of a network that can support the needs of business customers.

- *Low growth in residential video service.* The cable TV companies have been facing slow growth in their revenues from residential revenues. The growth in residential revenues is slightly higher than inflation.
- *Lack of national visibility.* Unlike the telephone companies, most cable TV companies lack not only visibility but also credibility with tele-com/datacom managers of large businesses.
- *High debt-equity ratio.* While the revenues of the cable TV industry exceeded $23 billion in 1993, the industry is heavily leveraged. This is reflected in the high debt-equity ratio of the major players, such as TCI with a debt-equity ratio of 5:1.
- *Proprietary systems.* Since its inception, the cable industry has operated primarily under proprietary systems. The cable systems of various cable TV operators lack common standards, making the interoperability among those systems difficult, if not impossible.

However, not all companies face every one of these challenges. Industry leaders such as TCI, Comcast, Continental, and Cox are well positioned.

8.1.2 Cable TV Goals

In response to the challenges facing them, the cable TV companies have defined the following goals:

- To generate new revenues by expanding the range of services they offer to their major customers (high schools, colleges, and universities) beyond basic TV services. New services include pay per view and video on demand.
- To create a low-cost, highly reliable, consolidated, and digital infrastructure to replace their predominantly analog cable TV systems.

To accomplish the first goal—the generation of new revenues—major cable TV companies have decided to enter the business market. They followed two approaches in entering the business market. The first approach has been to acquire AAPs. An example of these acquisitions is the purchase of Teleport Communications Group by TCI, Comcast, Continental, and Cox (most recently, Continental was purchased by U S WEST). The second approach has been the establishment of AAP subsidiaries. An example of this approach is Time Warner Communications, which was created as a AAP subsidiary of Time Warner.

The second goal of the cable TV companies—the creation of a highly reliable, low-cost, consolidated, and digital infrastructure—is being accomplished through cable network consolidation and the introduction of fiber-based systems such as HFC (see Chapter 1). The consolidation is being accomplished through the establishment of regional alliances through two financial mechanisms: joint ventures and lease-back arrangements.

8.1.3 The AAP Network Downtown

Through regional alliances, the cable TV companies and their AAP partners (or subsidiaries) have jointly built AAP–cable TV fiber backbones. These backbones, in some cases, originated in downtown areas where the AAPs operate. Figure 8.2 shows an example of an AAP network located in a downtown area. The network is based on a hybrid fiber ring–star topology. Nodes are configured as stars. The AAP runs fiber from digital cross-connect systems located in a central office to the customer floor. These digital cross-connect systems support private line services. AAPs are increasingly relying on ATM multiplexers located in the basement (specifically in common spaces) of the buildings of the customers they serve in offering business services.

8.1.4 The Regional AAP–Cable TV Networks

Through the regional alliances, the AAP networks located in downtown areas have been extended to the suburbs through the creation of regional networks (see Figure 8.3). This serves the following purposes:

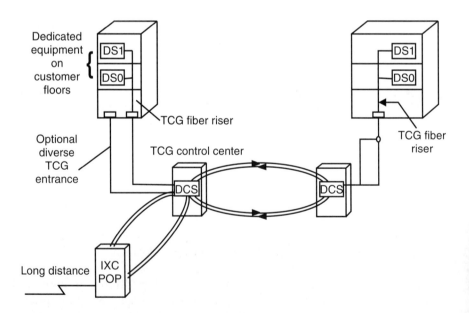

Figure 8.2 CAP network infrastructure.

- They enabled the AAPs to expand their networks from urban areas to reach industrial parks in the suburbs where business customers are located.
- They enabled cable TV companies sharing the same regional backbone with the AAP to serve their residential customers through their headends.
- In some cases, regional networks also included regional hubs, which enabled participating cable TV companies to further reduce their costs by providing them with a single point of regional advertisement and program insertion.

In some cases, the process was reversed: Regional networks were created first in the suburbs and then were extended to the urban areas, where the AAPs operate. It should be noted that the network sharing of cable TV companies is mostly limited to the physical transport and does yet not reach the application layer (i.e., cable TV companies use the regional network to serve their residential customers while the AAPs serve their business customers through services such as private lines, LAN interconnection services, and centrex services).

The joint regional cable TV–AAP networks enable the cable TV companies to reduce costs of deployment by sharing the costs of construction of the fiber backbones and reduce costs for regional advertisement and program insertion through regional hubs.

8.1.5 Future Cable TV Architecture

An improvement in the quality of the cable TV networks should be seen in the future (see Figure 8.4). A number of changes will improve the quality of the networks being built. The following are the likely quality improvements:

- *Systemwide deployment of ATM.* Initially, ATM was expected to be limited to the operating centers of the AAPs and to the regional hubs of the cable TV companies; the ultimate goal is to bring digital broadband connectivity closer to the residential and institutional users.
- *Emergence of server complexes as single repositories of video, audio, and data feeds.* These server complexes will be attached to the regional hubs and deployed as the single point of insertion of telecommunications services.

While regional networks represent the short-term objectives of the cable TV–AAP alliances, their long-term objectives are to build statewide networks and eventually their version of the NII. These plans will mature by the year 2000 or thereabouts. The timetable is driven by the fact that the process of negotiating such plans is long because it involves many players; also, the cost of implementing those plans is high, and cable TV companies lack the financial resources to implement them.

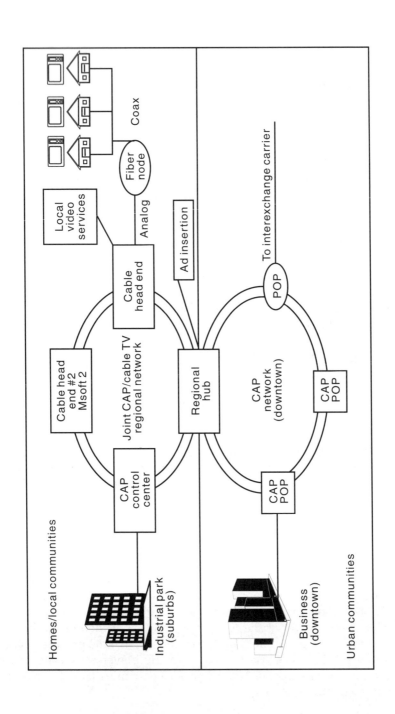

Figure 8.3 Current joint CAP–cable TV architectures.

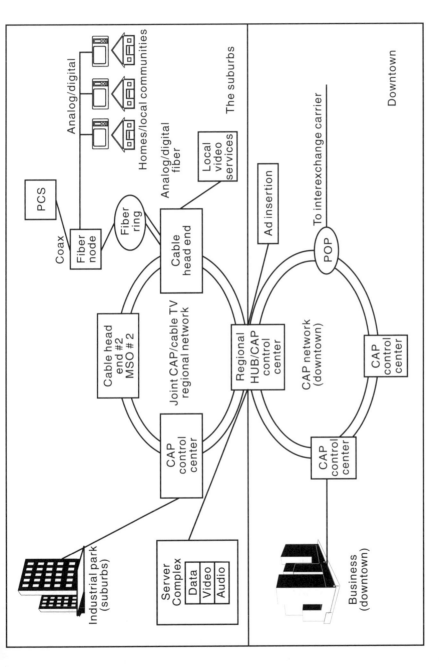

Figure 8.4 Emerging CAP–cable TV architectures.

A recent telephone company–AAP alliance is likely, however, to speed the process of creating portions of the NII by the cable TV companies. The alliance of Sprint, TCI, Cox, and Comcast should provide the opportunity for a technology infusion from Sprint and enable the members of the alliance to offer end-to-end solutions to their business and residential customers.

8.1.6 Cable TV Services and Markets

The regional networks built by the cable TV companies will enable them to introduce an expanding range of services. These services will be initially targeted at the key markets of the cable TV companies and the AAPs. Cable TV markets include the local communities, including K–12 schools and residences. AAP markets are predominantly financial service institutions. Following are brief descriptions of the current cable TV services that apply to those markets and to distance learning.

8.1.6.1 Educational TV Programming

Through cable TV, K–12 schools can access the following educational programs, among others, by subscribing to basic cable services:

- The Jones Computer program offers educational programming about computers, communication, multimedia, software, and related matters.
- The Horizons Cable Network provides cultural lectures, debates, and symposia.
- The Learning Channel provides general education and enrichment programming.

Students can also access a number of news channels such as CNN, CNBC, and C-Span.

8.1.6.2 Data-Cable Services

Data-over-cable networks are offered by cable TV network providers to support the needs of local communities in suburban areas. These local communities include residential customers, city government, court houses, and libraries.

Cable TV companies offer this service to take advantage of the extensive installed base of cable. As noted in Chapter 7, cable reaches 78% of all public elementary and secondary school districts and 62% of all public school locations. The data-cable solution can be used, for example, by K–12 students to access the Internet. It also enables residents to access corporate resources, for example, for telecommuting applications.

8.1.6.3 AAP Services

Cable TV companies offer a number of services through the AAPs that they own. These services are targeted primarily at the business market and include private-line services, at speeds ranging from to less than DS1 to SONET; ATM-based services, including NMLIS and FRS; and Centrex–single line dialtone service where competition is allowed.

8.1.7 The Activities of the Individual Cable TV Companies

This section provides more detailed descriptions of the some of the major cable TV companies, particularly their activities related to the distance learning market [1,2].

8.1.7.1 TCI

TCI is the largest cable TV company, with a presence in 49 states, operations in the 60 of the top 100 markets, and over 10.5 million subscribers. TCI strategic objectives in the education market are to enhance the image of the cable TV industry and to establish a foothold in the distance learning market. TCI offers a wide range of educational services, including a free cable hookup to 16,000 public and private schools and 500 hours per month of commercial free educational programming and data services.

To further support the distance learning needs of K–12 schools, TCI built a national teacher training center in Denver. The center assists and trains educators in effective applications of cable resources available for the classroom. In addition, students work in partnership with TCI to develop a learning and teaching environment using new technology tools.

TCI has aggressive plans to deploy fiber. TCI announced that by 1996 it expects 80% of its subscribers to be served by a fiber backbone system. To reduce the cost of deploying those fiber backbone networks, TCI has entered into joint ventures with other major cable TV companies, such as Cox and Comcast. Through those joint ventures, these cable TV partners have been cobuilding regional fiber backbones (e.g., in the San Francisco Bay Area). TCI also announced that it plans to spend $2 billion to upgrade its networks and support distribution of MPEG-2 video. To upgrade its networks, TCI is implementing, in some cases, a scheme that uses two coaxial cables to each feeder (the midrange goal, however, is HFC with one digitally modulated coaxial medium carrying digital video). One of the cables will initially be "dark," while the other primary cable will be configured with a downstream channel in the frequency range from 50 MHz to at least 750 MHz. When activated, the second cable will be midsplit with the "free" portions of 500 MHz allocated in each direction. TCI initially focused its efforts on upgrading its networks in 100 cities; then TCI was planning to expand its efforts to 400 cities by 1996. TCI plans to offer a range of services, including video on demand, interactive teleshopping, and interactive education programming, as well as data and voice data communications capabilities.

8.1.7.2   Time Warner

Time Warner was the second-largest cable TV company in the United States at press time, with over 7 million subscribers. In preparation for the introduction of video services on a wide scale and to support the distance learning market, Time Warner has been conducting a trial in Orlando, Florida. The trial is based on the full-service network (FSN) concept. The FSN is an interactive, digital, multimedia network that is intended to provide consumers with access to a wide range of services. The initial services included in the trial are video on demand, interactive shopping, and interactive games. Other potential services that could be added throughout the test phase include auto mail, more extensive retail shopping, video on demand services, news on demand, educational services, videophone, and personal communications services.

The FSN also provides participants with a navigational tool. The tool provides easy access to movies by letting them select services from a menu of choices. In addition, the tool provides participants with the ability to steer through a spatial interactive environment. The FSN delivers these capabilities to the homes by relying on ATM technology. The capabilities of the FSN differs from traditional cable architectures in one major way: With the current cable architecture, the same information is provided to all terminals and the home-based device chooses the information to be viewed by the consumer. In contrast, in FSN only the information that the user requests is routed to the consumer through the ATM switch [1,2].

The FSN was originally planned to begin with 4,000 households in April 1994, but it has been delayed several times and has encountered cost overruns. In early 1995, FSN began operations with Time Warner employees as its first end users; during 1995, the FSN was to be expanded from the initial video-on-demand offering to encompass distance learning, interactive games, and high data transfer. For the foreseeable future, the FSN will focus on residential customers only.

8.1.7.3   Continental Cablevision

Continental Cablevision is the third-largest cable TV concern; it has revenues in excess of $1 billion and serves about 3 million subscribers in 650 localities [1,2]. The company owns coaxial cable and fiber optic networks that support up to 50 channels of programming. The operator plans to continue to install several thousands of miles of fiber a year. In addition to basic and premium cable service offering, Continental Cablevision offers pay per view, locally produced programming, Cable in the Classroom, and Internet access. In late 1993, Continental Cablevision and Time Warner Cable, became the first cable TV companies to interconnect their networks to support end-to-end telephony services using PCS technology.

Continental Cablevision, along with Comcast Corp., Cox Cable, TCI, and Time Warner Entertainment, announced in early 1994 that they had agreed in principle to form a joint venture to develop new telecommunications services using digital video, fiber optics, and wireless technologies. The goal was to speed up the availability of competitive access telephone service, as well as develop new advanced services such as video telephony. The cable companies involved in this arrangement are already in the process of upgrading their existing fiber/coax networks with digital video transmission and two-way communications capabilities. Also in early 1994, Continental Cablevision announced that it was offering a 10-Mbps link over cable into the Internet for Cambridge customers. Continental has teamed up with Performance Systems International, an Internet access provider. The monthly cost of the service was estimated to be $70–$100, and the customer can access the network using a $150 Ethernet adapter. Recently, the company was acquired by U S WEST.

8.1.7.4 Comcast

Comcast is the fourth-largest cable TV provider. Like TCI, Comcast is a partial owner of TCG, which enables Comcast to generate revenues from the AAP market. Comcast is also providing cable TV services to schools and is testing and developing online (Internet) services to those schools. In Philadelphia, Comcast provided a school system with the ability to act as a program provider, producing content for two cable channels.

8.2 ALTERNATE ACCESS PROVIDERS

Following are brief descriptions of the activities of the two leading AAPs, Teleport Communications Group and Metropolitan Fiber Systems.

8.2.1 TCG

TCG is the largest AAP. The company offers a wide range of services to business customers. These services, which can be applied to distance learning applications, include:

- Private-line services, at speeds ranging from DS1 to SONET;

- ATM-based services, including CRS, FRS, and a NMLIS called LAN Interconnection Service (LIS);
- Centrex services as well as single-line dialtone, where competition is allowed.

As indicated earlier, TCG is part of a large alliance that includes three major cable TV companies and Sprint.

8.2.2 MFS

MFS is a relative newcomer to the telecommunications industry in general and to the distance learning industry in particular. Despite its relatively short life span, MFS has grown rapidly to become an increasingly important player in the industry, with revenues projected to exceed half a billion dollars in 1995. MFS played an important role in the distance learning industry by building the Iowa Communications network. MFS also offers a range of services to business customers. Those services, which can be applied to distance learning applications, include private lines, ATM, and centrex. Unlike other AAPs, which focus their attention on intraLATA services, MFS offers business customers connectivity at three levels: the intraLATA level, the interLATA level, and, increasingly, the international level. The MFS network currently spans and interconnects more than 30 cities in the United States.

8.3 SUMMARY

The cable TV companies are in the process of consolidating at two levels: the organizational level and the networking level. Organizational consolidation is taking place through a wide range of mergers and acquisitions. The net result of those consolidations will be the emergence of many fewer cable TV companies. At the networking level, cable TV companies are consolidating and upgrading their networks through the creation of regional networks. These regional networks will enable the cable TV companies to reduce their costs and to be in a better position to serve the distance learning (as well as the entertainment) needs of K–12 schools and residential customers. Cable TV companies, however, will not in be in position to directly serve the distance learning needs of business customers and large universities. That is because cable TV companies currently lack the credibility to serve business customers and because the networking needs of business customers extend beyond the regional level to encompass statewide, national, and international levels.

Cable TV companies, however, will be in a position to indirectly serve the networking needs of business customers (including their distance learning needs) through the AAPs. Cable TV companies have established alliances with

AAPs through ownership arrangements as well as through joint ventures. These alliances will be in a position to serve the networking needs of customers at the intraLATA level. The entrance of IXCs into the equation will enable those participating in the alliance to serve the networking needs of business customers at statewide, national, and international levels.

References

[1]   Minoli, D., *Video Dialtone Technology, Approaches, and Services: Digital Video over ADSL, HFC, FTTC, and ATM*, New York: McGraw-Hill, 1995.

[2]   Minoli, D., "Video Dialtone Overview," Datapro Report 1090, May 1995.

The Internet and
Distance Learning

9

The Internet is a major component of the NII. Some of the strong supporters of the Internet view the terms "NII" and "Internet" as synonymous, although others clearly disagree. The goal of this chapter is to explore the extent to which the Internet can be considered a short-term substitute for the NII, particularly in addressing the needs of the distance learning community. To accomplish that goal, this chapter describes the current and emerging Internet architecture; the services that those architectures support; the distance learning users and applications that can benefit from the services; the access speeds and services available to users; and the providers of those access services. The chapter concludes by analyzing the strengths and limitations of the Internet.

The Internet is an international network that, as of 1995, connected over 50,000 domains worldwide. Those domains serve over 5 million hosts. Traffic traversing the network at press time totaled 89 billion packets [1]. Estimates of the number of Internet users vary from 15 million to 20 million; half the Internet users reside in the United States. As a whole, the Internet continues to grow rapidly. The mix of Internet services and the relative traffic associated with each service continue to change as well. The services representing the majority of the traffic are FTP and WWW. Growth of the Internet traffic on the international level was expected to surpass that on the national level in 1996 [2].

9.1 THE INTERNET INFRASTRUCTURE

The Internet is a network infrastructure that was originally sponsored by a variety of federal agencies, including the NSF and the *Advanced Research Project* (ARPA). Until the first quarter of 1994, the Internet was made up of three tiers of networks: a DS1/DS3 backbone, regional networks, and local networks [2]. In

1994, the NSF received approval to triple the backbone speed to 155 Mbps b replacing the DS3 backbone with a backbone based on an ATM/SONET pla form (see Figure 9.1).

The Internet signed cooperative agreements with several suppliers of cor services for the next-generation network. The largest single award worth $5 million went to MCI Communications Corporation for the high-speed back bone. The domestic backbone connects NSF-funded supercomputer center throughout the country. The regional networks are midlevel networks centere around the NSF supercomputer sites. Those regional networks support loca or campus networks. In addition to the three tiers, a number of distinct com mercial networks (e.g., CompuServe) established gateways or bridges to th Internet.

9.2 THE ROLE OF THE INTERNET

Originally, the Internet was intended to be a network that would facilitate com munication and collaboration among researchers and educators in universitie government agencies, and industry. That, however, is no longer the exclusiv role of the Internet. Two growing roles of the Internet are supporting electroni commerce and providing distance learning solutions to K–12 schools.

9.3 INTERNET SERVICES

In its expanded role, the Internet supports an increasing number of service including the following [3]:

- Logon services;
- E-mail;
- File transfer;
- World Wide Web;
- Electronic whiteboards;
- Videoconferencing;
- Host-to-host communications;
- Directory services.

Each of these services is described next.

Logon services are offered by Internet providers through two types c TCP/IP–based protocols: Telnet and rlogin. The Telnet command has the ac vantage of being available on any host that allows remote login capability (Te net is supported by most Internet service providers [4]). Rlogin, a "Berkeley command," on the other hand, is not necessarily available on any host that a

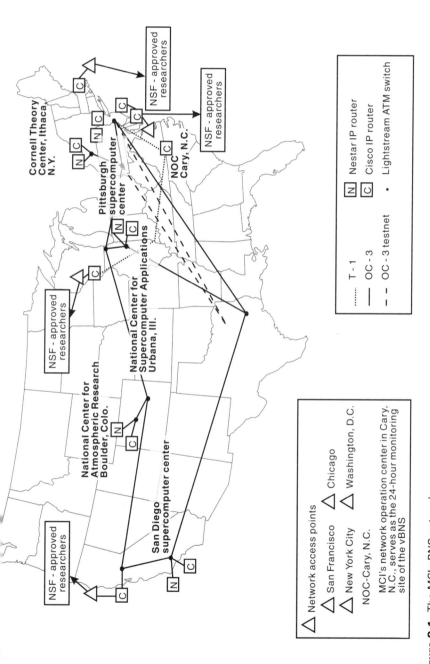

Figure 9.1 The MCI vBNS network.

lows remote login capability. The advantage of rlogin, however, is that it passe
on more information about the environment of the calling machine to the targe
host than Telnet does.

File transfer services offered by the Internet are based on TCP/IP trans
port/network protocols. FTP, riding on top of TCP/IP, is used to transfer file
between Internet hosts. Researchers at universities, corporations, and govern
ment agencies can use FTP to download as well as upload files to supercom
puter centers.

Anonymous FTP is a simple extension to the basic FTP protocol that er
ables students and researchers to create and share public archives with other
on the Internet. Currently, an extensive amount of information is available t
Internet participants through anonymous FTP, such as public domain or freel
available software, abstracts and full papers published by many researcher:
public domain books, and Internet standards. File transfer services are offere
by most Internet access providers.

Until 1995, e-mail was the most commonly used application on the Ir
ternet. (WWW now constitutes the most traffic on the Internet). E-mail is use
by researchers in universities, government agencies, and corporations as part o
their collaborative research process. E-mail is supported by a wide range of de
livery agents (or text editors) and delivery mechanisms. Examples of deliver
agents include Unix-based systems, such as Berkeley Mail program, and publi
mail programs, such as Pine and Mush. Example of e-mail delivery mecha
nisms are *simple mail transport protocol* (SMTP) and ITU-T X.400. SMTP is b
far the most common e-mail delivery mechanism and is used for transferrin
messages between different hosts. *Multipurpose Internet mail extensior*
(MIME) is a recent extension to basic Internet mail. While SMTP and X.40
cannot communicate directly, Internet providers make available gateways be
tween the delivery agents of both protocols. E-mail services are offered by mo.
Internet access providers.

The WWW is the most rapidly growing application on the Internet. Th
WWW is a system that links information stored anywhere on the Internet in
mesh of hyperlinks. These hyperlinks enable an educator or a student equippe
with a suitable client (e.g., Mosaic, WinWeb, or Netscape) to navigate through
distributed information resource by simply following pointers from one hype
media document to another. The hypermedia document may include te
sound, sound, still images, and video. The WWW makes information resourc
on the Internet accessible to users who are not experienced in UNIX-based con
mands. Consequently, the WWW makes the Internet a more useful tool to ne
groups of information users, such as K–12 students, teachers, and parent
These groups can access WWW from school buildings or their residences. Th
WWW has grown from 500 sites in 1994 to over 5,000 sites in 1995, and th
number of sites are still on the rise. The amount of WWW traffic on the princ
pal Internet backbones exceeded the amount of e-mail traffic. (The total amou

of WWW on the NSFnet backbone exceeded 1.3 terabytes in 1995, representing 8% of all Internet traffic.)

Electronic whiteboards allow a speaker's notes to be shared in real time with viewers over a wide area. This tool makes the dialog among researchers, educators, and students more interactive and more effective, enhancing the collaborative research and learning process.

Bulletinboard services are extensions of the e-mail service. This service provides a true public message, allowing users to send messages that can be accessed by other Internet members using similar bulletinboard systems. The best known bulletinboard service on the Internet is USENET; there are other bulletinboards, including gateways to commercial bulletinboards, such as Compuserve.

The Internet also offers access to online library catalogs for many libraries. To access the catalogs, students and researchers need to establish either a Telnet session or rely on a Gopher service. The Internet provides students and researchers with access to online databases such as Dialog, Dow Jones, and Lexis/Nexis. The Internet also provides users with access to the online library world, millions of public-domain shareware, and freeware programs.

Gopher services allow students and researchers to browse through information across the Internet without having to log in or know in advance where to look for information. The Gopher system offers information as a simple hierarchic system of menus and files.

Directory services provide users with the ability to locate information about other users, services, or service providers. Directory services are often divided into "white pages" and "yellow pages" services. The white pages service provides information about individual users, while the yellow pages service provides information about services and their providers. One of the earliest directory services is the WHOIS service, a basic user directory service originally conceived to track key network contacts for the early ARPA Internet.

The archie service catalogs the contents of hundreds of online file archives. The archie service collects the location information, names, and other details of files and indexes them in a database. Users can then contact an archie server and search this database for needed files. The archie service is accessible through a range of Internet services, such as Telnet and e-mail. The prototype archie service now tracks over 2 million filenames on over 1,200 sites around the world.

Videoconferencing is an emerging Internet service that enables K–12 students to establish low-end two-way video and audio connections with teachers and other students around the globe. This service originated as part of the Global Schoolhouse project. This service, when available on a large scale, will provide a measurable benefit to K–12 students, because it provides them with international exposure. It also reduces their physical isolation by enabling them to access external sources of knowledge, such as museums [5].

9.4   LOGICAL INTERNET CONNECTIONS

As noted, the Internet logical connections are predominantly based on the TCP/IP stack of protocols. The Internet will probably continue to be based on the TCP/IP protocol suite for the remainder of this decade, although some migration to IPv6 is possible. The IP layer is implemented on end systems as well as intermediary nodes. As a protocol, IP is responsible for internetworking multiple networks. These underlying networks may include LANs (e.g., Ethernet/IEEE 802.3, FDDI) or WANs (e.g., ATM).

TCP and *user datagram protocol* (UDP) are transport protocols that are implemented on end systems to packetize all application-layer exchanges into appropriate *protocol data units* (PDUs) and to provide transport services suitable for the applications. The transport layer of a protocol architecture controls the size of the PDUs, the number of PDUs that make up transmission, and the format of the data in the PDU. Several applications are supported by the transport layer, for example, SMTP, Telnet, and FTP (applications such as the Simple Network Management Protocol are supported by UDP.)

9.5   LOGICAL ACCESS TO THE INTERNET

The following two application-level solutions are available to users in accessing the Internet at the logical level:

- Time-sharing host solution;
- Full TCP/IP connection.

9.5.1   Time-Sharing Host Solution

The time-sharing host solution is more appropriate for the home and the K–12 networking environments. Through this solution, students and researchers can access the Internet (from their residences, dormitories, or schools) by obtaining an account on a time-sharing host that is located at the service provider's premises. In this case, the TCP/IP protocol stack resides on the time-sharing host [6].

To access the time-sharing host, the user needs to have a workstation and a modem. To establish a connection with the time-sharing host, the researcher or educator needs to establish a dialup connection to the terminal server attached to the Internet access provider's host. The dialup connection operates at speeds ranging from 2.4 Kbps to 14.4 Kbps. (While POTS is currently the predominant access method for users, ISDN is becoming increasingly available as an alternative method.) This dialup connection enables the user to dial into a terminal server that establishes a session with the time-sharing host. The host in turn establishes a connection across the Internet with the target host. A file transfer then occurs from the target host to the time-sharing host; the file is downloaded to the user's workstation through a serial file transfer protocol such as Kermit or Xmodem. Examples of time-sharing hosts are Internet public access sites, which are referred to as PDIALS. These public access sites exist predominantly in urban areas, such as New York City and San Francisco. They offer services, including e-mail, anonymous FTP services, remote login (Telnet), and remote navigators (e.g., Gopher). Charges for access range for $1 to $3 per hour (in addition, the user, who is using a modem, incurs the cost of telephone charges [7].)

Another example of public access sites are public access UNIX systems, sometimes referred to as "nixpubs." There are more of these types of public access sites than PDIALS. These types of sites support the user with a narrower range of services than those offered by PDIALS. Examples of these services include news groups, e-mail, and local source archives. Charges by these public access sites range from $5 to $15 per month. Another approach to accessing a time-sharing host is through 1-800-accessible services.

9.5.2 Full TCP/IP Connection

Full TCP/IP connection is another option available to Internet users. This solution applies mostly to corporate employees and to university students, researchers, and educators. The establishment of a full TCP/IP connection can be implemented in one of two ways: (a) a full TCP/IP suite may actually reside on the user's computer, or (b) a corporation may decide to establish an Internet node on the corporate side (in that case, the user accesses the Internet through a corporate access Internet node).

The first full TCP/IP connection option enables the user to run the TCP/IP protocol applications on a PC, including e-mail, bulletinboards (USENET), news programs, Telnet, FTP, and possibly Gopher. In that case, the user's workstation acts as a peer with other hosts on the Internet, which enables the user to move files directly from and to a hard disk, instead of going through the two-step process associated with accessing a time-sharing host. This approach also enables the user to access remote file servers, via *network file systems* (NFS) as apparent extensions to the file disk. Through that approach, the user can also support multiple sessions simultaneously in separate windows. The user can establish a connection between a TCP/IP-based computer and the Internet either on demand or on a full-time basis. TCP/IP services cost anywhere from $5 to $10 per hour.

The full TCP/IP interconnection option requires establishment of an Internet node on the corporate site. The user accesses the TCP/IP-based corporate host through a modem, a dialup line, and a serial line protocol. An example of the serial line protocols is the *serial line internetworking protocol* (SLIP). SLIP is a networking layer protocol that is widely supported by midlevel service providers and campus computing centers.

9.6 INTERNET ACCESS SPEEDS

A number of physical access methods and speeds are available to the educator and researchers. Access methods can be divided into two categories: low-speed access methods and high-speed access methods.

Low-speed access methods include dialup connections, X.25 dialup, toll-free dialup, and ISDN. Dialup connections are offered by the major Internet service providers. The connections range in speed from 9.6 Kbps to 28.8 Kbps. Toll-free dialup is offered by the major Internet service providers. These connections also range in speed from 9.6 Kbps to 28.8 Kbps [6]. ISDN access to the Internet is increasingly available. For example, Performance Systems International began providing ISDN BRI in 30 cities [8]. These dialup solutions are adequate for students who need to communicate with other students or teachers from their homes or dormitories. These solutions are also adequate for telecommuters who would like to access information resources or establish corporate connection.

High-speed access methods include two categories: fast packet services such as SMDS, ATM, and FRS, and cable TV solutions, such as data/cable modems. The fast packet access solutions are offered by an increasing number of access service providers, while the data/cable solutions are offered by cable TV providers, such as TCI. High-speed solutions are more appropriate for universities and corporations in support of applications, such as access to super computers. As discussed, the Internet infrastructure is migrating to a ATM/SONET infrastructure. When fully implemented, this infrastructure will provide higher performance to corporate users and universities.

9.7 INTERNET SERVICE PROVIDERS

Internet services are offered by a number of providers, which can be categorized as national Internet providers and regional/local Internet providers. The two groups of providers differ in terms of their geographic presence, the number of services they support, and the network solutions they offer. Differences also exist among the providers who belong to the same group.

National Internet service providers operate in multiple cities across the United States. These providers provide participating members with access (gateways) to other domestic and international networks. They include the Advanced Network and Services, Inc. (ANS), California Education and Research

Federation Network (Cerfnet), Performance Systems International, UUNet Technologies, Inc., and Sprint Corporation.

Regional Internet providers include the San Francisco Bay Area Research Network (BARRNet), Committee on Institutional Cooperation Network (CICNet), and Southwestern States Network (WestNet), all of which operate in the western region of the United States. Global Enterprise Services, Inc./Northeast Regional Network (JvNCnet) and New England Academic and Research Network (NEARNET) operate in the Northeast. Committee on Institutional Cooperation Network (CICNett) operates in the Midwest. Southeastern Universities Research Association Network (SURANet) operates in the Southeast.

The number of the Internet service providers is on the rise, because the Internet represents an increasingly attractive business opportunity to network service providers and to hardware and software suppliers. Internet market revenues (1995) can be allocated to the following segments [9]:

- The hardware sale segment, worth $476 million, includes revenues from the sale of hardware, such as routers, hubs, and servers.
- The expertise segment, worth $59 million, includes revenues that can be generated from supporting users in connecting to the Internet, setting up e-mail servers, and designing "home pages" for the Internet's WWW.
- The access segment, worth $135 million, includes revenues generated by the Internet providers, such as UUNet.
- The software segment, worth $160 million, includes revenues generated by the software developers that provide Internet navigational tools.
- The content provider segment, worth $6 million, includes universities and the library of Congress.
- The commerce (e.g., ordering lunch through the Internet) segment is currently worth $6 million.

Recognizing the large business potential of Internet services, several groups of network service providers have entered the Internet access market. These groups include the IXCs, the local exchange carriers, and cable TV providers.

Among the IXCs, Sprint was the first IXC to offer commercial TCP/IP services and gateways to the Internet through SprintLink. In 1994, Sprint expanded its SprintLink Gateway into Global SprintLink. As part of its services, Sprint is responsible for the operation and maintenance of network routing, as well as maintenance of the physical network. SprintLink also offers a whole array of navigational tools and database extensions. In addition, SprintLink Plus offers enhancements to Sprint's basic offerings, including firewalls and security features.

AT&T is another IXC that offers Internet-related services, including dialup Internet access and directory services. AT&T is the official directory and service provider for the InterNIC, which is the body responsible for registering all Internet names and addresses. AT&T offers three groups of Internet connection offerings, including Easylink and the Interspan Family of Services. EasyLink provides a main gateway to the Internet that enables subscribers to exchange e-mail. The InterSpan Family of Services includes two Internet-related offerings: dialup access at 14.4 Kbps and FRS.

In addition to the IXCs, the local exchange carriers have announced plans to offer Internet services. For example, Pacific Bell announced in early 1995 the availability of Internet services. In addition to the local exchange carriers, cable TV companies are also increasingly offering Internet services. AAPs also offer access.

9.8   MULTIMEDIA OVER THE INTERNET

The *Internet Engineering Task Force* (IETF) is now working on a new set of routing and transmission standards to improve multimedia conferencing over the Internet. Bandwidth-intensive applications require both high-capacity transmission and low-latency protocols to be effective. The IETF standards, still in draft at print time, are aimed at improving support for networked multimedia through bandwidth reservation and more effective use of multicast capabilities [10].

Sending duplicate messages or setting up broadcast video–data collaboration conferences to multiple remote locations requires multicasting. The IETF standard, *distance vector multicast routing protocol* (DVMRP) is a multicast routing algorithm used in the Internet multicast backbone network to connectivity (e.g., video) to workstations that support IP multicast (RFC 1112, 1989). DVMRP protocol operates by flooding all available routes. Protocols now under development are more efficient. *Protocol-independent multicast* (PIM) uses "rendezvous points" for registration of the senders and receivers of multimedia multicast traffic. Being protocol independent, it works with any unicast routing protocol. A standard also under development is real-time transport protocol, which uses time stamps and content identification labels on multimedia data units to ensure reassembly at the receiving end.

Streaming II (ST-II) is a connection-oriented protocol completed in 1991 (RFC 1190, also known as IP Version 5) that has been deployed in the Defense Department (but not much elsewhere). This protocol lets the originator control the bandwidth and latency in a multicast videoconference. A revised version, called ST-II+, allows for faster connection setup and the ability of allowing receivers as well as senders to join a multicast group without needing a central administrative point; it was under development at press time. Vendors such as Wellfleet Communications, BBN, IBM, and Syzygy were early supporters. ST-II+ is not backward compatible with ST-II.

The *ReSerVation protocol* (RSVP) offers a way to support bandwidth allocation on connectionless networks. It decentralizes control over bandwidth allocations, letting each end system decide how much bandwidth is required to participate in a multimedia conference. As in ST-II+, users can join the conference without a centralized administrative function.

These protocols can be used for Internet-based multimedia conferences in support of distance learning applications.

9.9 STRENGTHS AND LIMITATIONS OF THE INTERNET

Although it is absolutely clear that, in the end, the NII will comprise a collection of infrastructures from various industry segments, some observers (incorrectly) view the Internet as the "true" NII. They attribute their belief to the following strengths of the Internet:

- *Support by network service providers.* Access to the Internet is increasingly supported by not only the traditional Internet providers but also the LECs and the IXCs.
- *Number of people reached.* Estimates of the number of people interconnected by the Internet vary from 15 million to 20 million.
- *High-performance backbone.* As indicated, the NSF backbone is migrating from a DS3 platform to a higher-performance network, based on ATM and SONET.
- *Expanding range of applications.* Traditionally, the Internet has supported data networks, which serve the research and collaboration needs of researchers and educators employed by universities, corporations, and government agencies. The role of the Internet is expanding to encompass supporting entry-level two-way video and audio applications.
- *Affordability.* The Internet remains one of the least costly approaches to provide interconnection, delivering on the NII promise of an affordable networking solution. Internet cost to users range from $1 to $10 per hour in addition to telephone charges.
- *Extensive information resources.* This represents a source of strength of the Internet as an application solution to distance learners. Through the Internet, the distance learning community can access hundreds of libraries around the world, as well as library catalogs and full text-delivery services. The Internet also enables distance learners to access a wide range of government agency information. In addition, the Internet provides participants with a wide range of public domain software, freeware documents, databases, images, and other files.
- *Ease of use.* New services, such as WWW, make the Internet easy to access and to use.

All these factors make the Internet an important component of the NII, but it cannot be considered a complete substitute for the NII for the following reasons:

- *Lack of support for rural communities.* While it is widely available, the Internet has not yet reached rural communities with local access nodes (remote dialin, either through a 10-digit or an 800 number is, however, an option). Rural communities are one of the key drivers behind the need for a ubiquitous NII.
- *Lack of information filters.* A range of information resources on the Internet are not of the highest (or any) educational, social, or moral value [11].
- *The Internet as a public forum.* Like talk radio, the Internet has become a public forum, where ideas are generated, discussed, and debated. Some of these ideas, however, spread negativism and cynicism.
- *No guaranteed performance.* The distance learning community needs to consider that it cannot be guaranteed a certain throughput across the Internet nor a consistent reliability level. That is because the Internet, while serving thousands of organizations and millions of individuals, lacks any mechanism for reserving bandwidth.
- *Lack of antiviral software.* Internet e-mail or downloaded files have been known to contain viruses, which cannot be eliminated without antiviral software on every machine in the network [12].
- *Junk e-mail.* Institutions connected to the Internet can be flooded with useless and unwanted e-mail.
- *Security issues.* The security weaknesses of the Internet limit its usefulness to interorganizational communications, including those associated with distance learning solutions.

Several solutions are becoming available to corporations that want to improve the security of Internet access [13]:

- *Screening router firewall.* This solution involves using routers configured using native access rules to filter packets at the IP port level. The benefit of this solution is that it may not require additional investment on the part of the corporation since the router may be a part of the installed base. The solution, however, lacks customization and audit capabilities.
- *Application-level firewalls.* This solution provides a higher degree of security than a router firewall. The implementation of this method is accomplished by forcing users to access the system only via application-level prompts. The benefit of this solution is its audit capability. Products that support this solution, however, can be expensive, and they are relatively new. Another limitation of this solution is that it requires administrative commitment.

References

[1] "International Internet Growth Poised to Overtake Domestic in 1995," *Telecommunications*, Dec. 1994, p. 12.

[2] Masud, S., "Nsfnet Makes Leap to ATM/SONET," *Government Computer News*, Mar. 7, 1994, p. 39.

[3] Eldib, O., and D. Minoli, *Telecommuting*, Norwood, MA: Artech House, 1995, pp. 93–97.

[4] Bosco, P. D., and H. V. Braun, "The NSFNET T1/T3 Network, Backbone of Emerging Information Infrastructure Celebrates 5 Years of Extraordinary Growth (1988–1993)," *Connexions*, Sept. 1993, p. 2.

[5] Serf, V. G., "The Internet's 25th Anniversary. We Think!" *Telecommunications*, Jan. 1995, p. 23.

[6] "Internet Primer," Technical notes, p. 36.

[7] Dern, D. P., "Start Here: How and Why to Join the Internet and Get Going," *Internet World*, Sept./Oct. 1993, p. 68.

[8] "ISDN Access Provides Cheaper Internet Links," *Computerworld*, Nov. 22, 1993, p. 57.

[9] Flynn, L. "Internet Server Takes a Big Step," *New York Times*, Feb. 5, 1995, p. F10.

[10] Minoli, D., and R. Keinath, *Distributed Multimedia Through Broadband Communication Services*, Norwood, MA: Artech House, 1994.

[11] "Cyberporn—Porn on the Internet," *Time*, July 3, 1995, pp. 38 ff.

[12] Layland, R., "A Gateway to Internet Health and Happiness," *Data Communications*, Sept. 21, 1994.

[13] Minoli, D., *Web, Internet, and Related Technologies and Services for Corporate Environments*, New York: McGraw Hill, 1996.

Bibliography

Chapin, A. L., "The State of the Internet: 1995," *Telecommunications*, Jan. 1995, p. 24.

Dougherty, D., and R. Koman, *The Mosaic Handbook*, O'Reilly & Associates, 1994.

Engst, A., *The Internet Starter Kit*, 2d ed. Hayden Books, 1994.

Hahn, H., and R. Stout, *The Internet Yellow Pages*, 2d Ed., Osborne McGraw-Hill, 1995.

LaQuey, T., *The Internet Companion: A Beginner's Guide to Global Networking*, Reading, MA: Addison-Wesley, 1994.

Turlington, S., *Walking the World Wide Web*, Ventana Press, 1995.

Case Study: Implementing High-Performance IDL Systems in K–12

10

10.1 IDL: WHAT IS IT AND WHY USE IT?

IDL is an evolving paradigm of instruction and learning that attempts to overcome both distance and time constraints found in traditional classroom learning. It is a set of technologies that can allow for a more equitable distribution of resources, as well as a more personalized learning experience. By implementing IDL, a school system might:

- Offer Japanese to a class of 20 students located at four different middle schools;
- Offer calculus to students in the inner city at several different high schools that otherwise could not justify a calculus teacher;
- Allow students to work ahead of or behind their grade in school in areas of particular importance without the complications of changing a grade level;
- Allow students to undertake collaborative work with other students and teachers that would otherwise not have been possible.

This case study will focus on meeting the needs of K–12 students, since building IDL systems for this group of end users represents a significant technical challenge. Live IDL between multiple sites as well as the client/server model of host-based distance learning are considered. University and corporate applications that strive for a richer, more intimate live IDL experience can fol-

---

This chapter was provided in its entirety by Marc P. Pfeiffer, director of product marketing, Newbridge Networks (marc_pfeiffer@qmail.newbridge.com). Since 1988 he has been involved in the design and implementation of numerous regional and statewide IDL networks. His current responsibilities include driving the development of client/server-based video applications that run over ATM LANs and WANs.

low these methods for room, network, and system design. The architectures covered are limited to terrestrial telephony and data networking technologies Given that technological developments in all aspects of IDL continue unabated this case study considers the state of the art as of 1996.

10.1.1   Striving to Meet Expectations

It is equally as important for implementors of IDL to understand what IDL is not. IDL will probably never replace the individual attention a teacher can give to a student when they are both in the same classroom. It does not today offer a sense of intimacy that one experiences around a table discussing the finer points of Chaucer in an English seminar. But these are precisely the high-quality learning experiences IDL should strive to simulate with appropriate technologies and high-quality system design. The big challenge in IDL design is to offer a true sense of *distance* learning, not *distant* learning. As IDL continues to evolve, maybe some time in the future it will indeed seem as if our peers and the teacher are in the same classroom with us when they are actually many miles away.

In an effort to build the appropriate IDL system for a given learning requirement, it is important to establish a basic set of objectives that the system must meet. For example, what audience is the implementor trying to reach K–12, university, corporate? To what degree will the mode of instruction be lecturing versus interactive discussion? What is the desired degree of interactivity among the remote sites as opposed to just between each site and the teacher?

Given that the classroom experience represents a far greater percentage of the total learning experience among younger students, it is in K–12 that IDL systems are put to the greatest test. By and large, the top 20% of students will learn no matter what the conditions. It is the remaining 80% that the implementor must strive to reach with a meaningful educational experience. And while IDL is often viewed as an extension of traditional video teleconferencing in the university and the corporate world, that model generally does not meet the objectives of classroom teaching in K–12. Technical specifications on picture quality, motion artifacts, audio quality, and end-to-end latency are going to have to be much tighter when dealing with a younger audience. If it looks different than what is seen on TV in the home, chances are you will lose this group of end users. If the VJs on MTV make eye contact on television, then so must the teacher if IDL is going to look natural. If the music videos are free of motion artifacts, then that will be a requirement of IDL as well. The more natural an IDL system looks to the students and the teachers who use it, the more transparent and uninhibiting the technology will become.

Done right, IDL can be an effective and powerful learning tool that not only augments classroom learning but has the potential to bring a whole range of new learning experiences into the schools. Those experiences might include a live video class trip aboard the space shuttle, a lecture from a famous guest speaker who answers students' questions at the end, or face-to-face contact with students of entirely different backgrounds. The key to making those things happen is to be sure from the start that technology enhances rather hinders the experience. Most school systems will have one shot at building and proving in distance learning. Projects often start as a pilot programs before significant funds are spent. Proving in the best performing system the schools can afford will go a long way to ensuring a reasonable return on an IDL investment.

It should also be made clear that teachers often feel that technology is thrust on them by people who possess far greater technical literacy than they do. In general, teachers have had very little preparation for new technologies brought into the classroom, and IDL has been no exception. Teaching is not the same over an IDL network, and teachers will have to revise lesson plans and teaching styles to make the most of the medium. That will require a good deal of preparation on the part of the teacher as well as an IDL coordinator to be sure all aspects of the IDL session run smoothly. Furthermore, union contracts and rules imposed by state and local boards of education will govern an IDL session just like any other class. That includes class size, number of hours of instruction in a specific subject, and demonstration of proficiency.

10.1.2 Two IDL Paradigms: Live and Stored

Before the implementor goes ahead with the design of an IDL system, it is important to establish a framework for understanding the different modes of digital video communications. Given an appropriate compression scheme, digital video can be shown live over an IDL network, or it can be stored for on-demand viewing. Once on a computer, digital video can be combined with other media such as still image and text to form multimedia source material. The live mode is the one most often considered in the planning of an IDL network. This usually involves connecting students and teachers over a network for a class session that would otherwise not have been possible. The live paradigm will be covered in a fair amount of detail, including room design, video processing, transmission and switching, and session management.

As host computers, video servers, and data networks become more powerful, host-based learning on demand will become an important component of IDL networks. Although the client/server component of IDL is just beginning to be developed, it is here that the technology can make a personal difference to students and teachers alike. That should be clear to anyone who has watched kids interact with CD-ROMs on multimedia computers.

At the same time, just introducing CD-ROMs to the class can be problematic for several reasons. First, since CD-ROMs are limited to single-access use, every student who wants to use a particular disk must obtain a copy. That requires the distribution and management of many copies of the same CD-ROM throughout the school system. Host-based IDL, on the other hand, can be inherently multiaccess by relying on a client/server network. It is envisioned that the stored component of IDL will include file servers for computer-based learning and video servers for multimedia. For example, multiple access of online material can be made available to schools over a high-performance data network based on ATM. The network architecture will support online updating of the material as well, which is another significant improvement over the more permanent CD-ROM. The fact that on-demand subject matter can be stored on servers will eventually allow for a good deal of source material to be developed by students and teachers. Programs are already available costing under a thousand dollars that offer fairly functional video editing and effects for producing multimedia material. Furthermore, the on-demand component of IDL should also help create a large and diverse market for online educational source material.

Another key component of network-based learning is access to the Internet. Though not covered in this case study, the Internet represents one of the most important online resources available to schools today. If funding for IDL is limited, consider Internet access and e-mail as an excellent first step to bringing remote learning resources to the schools.

10.2 FOUR COMPONENTS OF LIVE IDL

Live IDL can be broken up into several components that must be tightly integrated to achieve a seamless network. Those components include room design, video processing, transmission and switching, session management, and network control. Tradeoffs made in any one of those areas can greatly affect the overall performance and operation of the entire system. In deciding what tradeoffs are appropriate, it is key for implementors of IDL networks to establish a set of objectives that the system must meet, including functionality, quality, and cost. For example, establishing system design objectives might include the following:

- The degree to which the IDL session is to simulate the intimacy of teaching in a single classroom;
- The degree of interactivity between students and teachers, as well as among students in different classrooms;
- The ratio of lecturing versus interactive discussions, for example, 90:10 or 50:50;
- The maximum number of participating sites in a general session;
- Alternative modes of operation to address different needs that arise, such as general broadcast to all sites or larger interactive sessions;

- Alternative after-school uses that may help defray some of the costs of operation.

In answering those questions, the implementor will establish a set of baseline operating parameters that must be met by an optimization of all aspects of system design. The importance of establishing system design objectives cannot be overstated. A great deal of time, energy, and money will go into bringing any IDL network on line. It is key that once the system is up and running that it indeed fulfills the objectives for which it was designed.

10.3 ROOM DESIGN

The layout of the classroom is paramount to the success of any K–12 IDL network. Classroom design will determine the way a teacher teaches a remote session, the number of participants in a class, and their ability to interact with teachers and students at remote sites. These parameters can greatly influence the overall quality of the learning experience. When budgeting for IDL classrooms is being considered, it would be better to invest in fewer rooms "done right" than in a large number of rooms that fall short in terms of overall functionality. Two key areas of high-quality room design are excellent audio and an appropriate seating layout to achieve good visibility of both cameras and monitors.

High-quality audio starts with low room noise. Cumulative noise is a major deterrent to a productive distance learning session. In a multisite IDL session, the overall ambient noise is the sum of the noise added by each classroom plus any transmission impairments introduced by audio processing in the network. Some key contributors to noise are mechanical systems such as heating and air conditioning, a "live" room with echoes bouncing off hard surfaces, street noise from an open window, and a misaligned room audio system.

Low ambient room noise greatly enhances the overall learning environment and can be achieved by following relatively straightforward guidelines. If possible, choose a room that has no windows. That helps to reduce noise as well as maintain a constant lighting condition. It is also a good idea as far as security is concerned, since these rooms will contain some of the most desirable equipment found in a school. Low ceilings are also a good idea, to keep echoes to a minimum. The "liveness" of a room can be greatly reduced by having carpeting on the floor and sound-absorbing wall treatment, which can be as simple as covering several 4-foot by 8-foot pieces of plywood with thick pile carpet and hanging them on the walls. (Every attempt should be made to match the overall aesthetics of the room so the treatments themselves are not distracting.)

The real trick in a closed room is to provide adequate heating, ventilation, and air conditioning without introducing the mechanical noise that usually accompanies such systems. That may indeed prove problematic, but if it can at all be addressed, minimizing mechanical room noise will go a long way to reducing overall ambient noise.

Lighting on both the teacher and the students is important as far as the remote sites getting a good look at each other. A heavy shadow or a bright wash-out of the teacher may become an impairment to foreign language instruction, for example, where actually seeing the articulation is key to the learning process. As mentioned, a room without windows allows for a stable and controllable lighting condition. One further step to ensure the best possible image quality is to select lighting with a temperature balance that matches the response curve of the video cameras. That will ensure that skin tones are natural and other subtle color information will not be lost.

10.3.1 Video and Audio Components

Depending on the system design objectives set out in the first phase of an IDL project, there will be a certain number of cameras and monitors in each classroom. In most cases, there is a minimum of three cameras: one in the back of the room for the teacher's shot, one in the front of the room for a shot of the students, and one overhead graphics camera. The graphics camera is usually built into a presentation stand that is part of a teacher's podium. The two room cameras should be of the three-chip, *charged couple device* (CCD) variety with reasonably high resolution for good contrast. Each camera should be equipped with a lens of appropriate focal length to provide the best possible shot of the class or the instructor. There is a broad range of industrial cameras and lenses that will work well in an IDL classroom. It is important to work with a video systems integrator to determine the best combination of camera and lens for a specific classroom layout.

The objective of camera placement is to coincide wherever possible with monitor placement. The reason for that is to strive for a modicum of eye contact between the presenter and the students at the remote site. Since teacher and students will be looking at the monitors to see the remote sites, a good location for room cameras is in close to or between these monitors. This is equally as important for the teacher's shot as for the students', since nothing diminishes a sense of intimacy more than lack of eye contact. There's nothing worse than seeing the top of an instructor's head who is looking down into a desk-mounted monitor to answer an incoming question. Figure 10.1 shows camera and monitor placement for a typical four-site continuous-view configuration involving eight monitors and two cameras plus a graphics overhead camera.

The next critical component to live IDL is audio, which is broken up into two functions. A room pickup system consists of student mikes, the teacher's

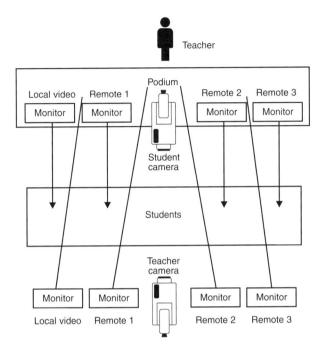

Figure 10.1 Full continuous-view room layout.

mike (which may be wireless), noise cancellation, an equalizer, and an input mixer. There are several room pickup systems designed for high-quality teleconferencing that work well in IDL applications. Mikes should be distributed one for every two students in the room and preferably installed on the desks. The return audio system is made up of an audio amplifier and room speakers. This system should allow everyone in the room to hear the incoming audio feeds without feeding back through the pickup system.

10.3.2 Five Rules for Room Design

The five rules of high-quality room design are as follows:

- The signal will never be better than when it leaves the classroom, so cutting corners there will affect everything else down the line.
- Audio impairments will kill an IDL session long before video impairments will.
- Noise is the audio signal's worst enemy, and ambient noise adds up from every participating site.
- No eye contact, no intimacy.

- Room layout should be designed to support a high level of interactivity with an intuitive layout that is easy to use.

10.4 VIDEO PROCESSING

All networked digital video is compressed to some degree. That is because digital video is a bandwidth-intensive signal, requiring roughly 150 Mbps when transmitted uncompressed. Compression ratios of 10:1 to 2,000:1 are common and highly applications specific. Compression is used not only to reduce bandwidth for transmission but also to reduce memory cost for on-demand applications that are storage based.

When it comes to video processing tradeoffs in live IDL, numerous parameters must be considered. Those parameters can be grouped into three areas: video information content, including resolution and color information; motion-handling capability; and latency due to signal processing. In assessing image quality tradeoffs of these key parameters, it is essential to refer to the system design objectives set at the beginning of the project.

10.4.1 Image Quality Requirements

The problem with K–12 IDL networks is that the end users, namely, the students, are unabashed video-quality experts. Image quality that is anything less than that seen in the home can become an impairment to learning. This is less of an issue as the maturity level of the students goes up. However, for a general-purpose IDL system addressing the needs of K–12, the video will have to be entertainment quality. Figure 10.2 shows the relative image quality required with respect to the age and the maturity of the student.

Because IDL is a full duplex communications medium, overall system latency, which includes the time required for video compression, transmission, and video decompression, must be kept to under 250 ms. Anything greater than that will reduce its overall effectiveness as an interactive communications tool. The greater the compression, the more latency will limit the degree of interactivity that can be supported over the network. Spontaneous dialog is lost. Latency generally is not a problem in a broadcast-only scenario, since there is no return communications channel involved. Thus, broadcast applications geared toward entertainment can trade off computationally intensive compression and its resulting latency for better picture quality at a given bit rate. That is how a digital-based entertainment service provider generally uses MPEG-2 compression.

To summarize, the image quality for IDL applications geared toward K–12 must be at least as good as that seen in the home. This includes NTSC resolution at 30 frames per second, good color saturation, no motion artifacts, and

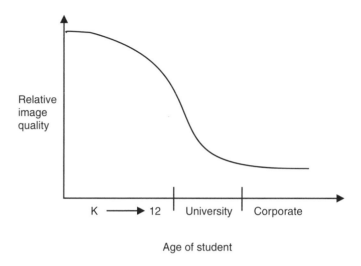

Age of student

Figure 10.2 Image quality requirements.

good audio. And because IDL is used as a full duplex communications channel, overall system latency must be kept to a minimum.

10.4.2 Video-Compression Options

Given the tight constraints on video quality called for in live K–12 IDL applications, the range of acceptable compression options is fairly limited. Motion artifacts and latency due to high rates of compression are the primary limiting factors. Motion-handling capability is important not so much for the everyday teaching that goes over the network but for the distribution of video source material, such as live news feeds, cable television, a downlinked satellite feed, or videotaped educational source material.

Compression ratios ranging from 5:1 to about 30:1 for live applications at full NTSC resolution are currently required for entertainment-quality video with low latency. Those ratios can be achieved with motion-JPEG compression, and proprietary compression algorithms used for broadcast-quality video transmission, generally referred to as *back hauls*. (In the future, low-latency MPEG-2 compression using I and P frames only may eventually be adequate for live IDL.) Actual bit rates for live IDL applications range anywhere from 5 Mbps through 45 Mbps. That may seem like a lot of bandwidth, but it is the only way to maintain high image quality, good motion-handling capability, and low latency in live, multipoint video applications. Stored applications can undergo a good deal more compression with acceptable results. Table 10.1 gives a summary of hardware video-compression alternatives, including bit rates and compression ratios.

Table 10.1
Hardware Compression Overview

Format	Bit Rate	Quality Level	Compression Ratio
D-1 digital component video (CCIR 601, 4:2:2)	270 Mbps	Studio	1:1
D-2 digital composite video	143 Mbps	Broadcast	1:1
DS-3 proprietary compression	45 Mbps	High-quality distribution	4:1
Motion-JPEG compression	6–40 Mbps	High-quality distribution	10:1
MPEG-1,2 compression	1.5–6. Mbps	Entertainment	50:1
H.320/H.361 ITU-T standard	128K–1.5 Mbps	Conferencing	1000:1

Certain key parameters should be kept in mind during the evaluation of compression scheme:

- Does it provide adequate image quality and motion-handling capability to meet the stated objectives of the IDL project?
- Is end-to-end latency low enough to support a high level of interactivity between several sites?
- Are encoding and decoding fairly symmetrical as far as latency and cost are concerned?
- Is the capital cost of video compression balanced off against the recurring operating charges for digital bandwidth?

It should also be made clear that advances in video-compression techniques and implementations press on unabated. Cost-performance ratios will continue to go down, as will (hopefully) the cost of digital bandwidth. However, it is important to choose a proven technology that is available today and that will meet the stated objectives of the IDL network.

10.5 TRANSMISSION AND SWITCHING ARCHITECTURES

10.5.1 Broadcast, Session Switching, and Continuous View

After room design and picture quality, the most important technical decision the implementor must make is system architecture. IDL network design will ave a direct effect on the types of sessions supported, the amount of interactiv-y that can be expected, and the overall cost of the system. Typically, network esigns include broadcast, session switching (also known as scanning), and ontinuous view. Broadcast implies a one-way point-to-multipoint videocon-rence with audio channels possible in the return path. That is the most lim-ed mode of distance learning as far as interactivity is concerned. It is primarily eared to a lecture style of instruction at the college level, with little if any in-raction between remote sites. Terrestrial broadcast distance learning has been idely implemented over microwave in the ITFS instructional television band t 2.5 GHz. Broadcast alone is rarely used as a primary mode of interactive in-truction for K–12 applications because of the limited return-channel support.

As a result, nearly all terrestrial IDL networks designed for K–12 can be escribed as continuous-view,[1] session-switched, or a combination of the two. ull continuous view was the first architecture adopted for IDL networks pri-arily as an outgrowth of the cable TV technology that was originally used to nplement these systems. An analog RF broadband bus architecture connected ll schools by way of the local cable TV coax plant. When a classroom wanted participate in an IDL session, it simply tuned in the other remote sites on the etwork by changing channels on a television set. Those networks were opti-ized for four participating sites per session (one instructor site plus three re-ote sites), primarily to make room designs manageable. Each room required ight monitors: a set of four for the teacher and four for the students (one local onitor plus three monitors for each of the remote sites; see Figure 10.1 for full ontinuous-view room layout). This approach had several drawbacks, includ-g limited reach, limited number of participating sites, no support for data, nd no network management. It did work, however, and was used successfully y numerous school districts and cable companies in support of IDL in the 980s. Today, however, nearly all new IDL networks are going in digital.

The strongest argument for full continuous view is that it offers the great-st amount of intimacy between students and teacher. With a full continuous-iew architecture, the teacher can walk into the class and immediately see and ear all students at the remote sites throughout the session. Students see and ear the teacher as well as all other students at the remote sites. The level of teractivity possible approaches that of a regular classroom. Full continuous iew offers the most options in teaching methods, and when given the choice, it generally preferred by pedagogues who teach K–12. The disadvantage of the rchitecture is that it is bandwidth intensive and requires a fairly sophisticated nd potentially expensive room design to achieve the best results. Figure 10.3 llustrates a layout of a full continuous-view network.

An alternative to full continuous view is to offer some kind of switching echanism for the teacher to select the video coming back to the instructor site.

. Referred to as *continuous presence* in previous chapters.

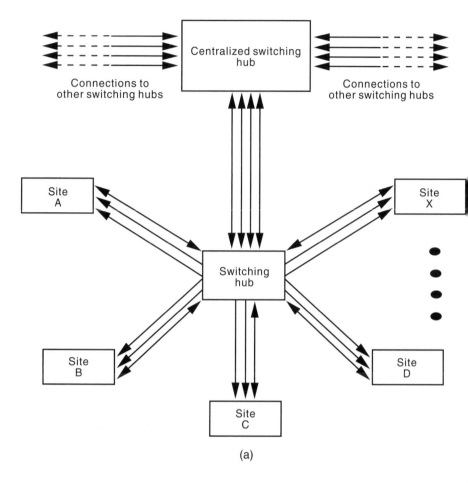

Figure 10.3   (a) Full continuous-view network, and (b) full continuous-view network in detail

This can be done in numerous ways and even be automated. The advantage of session-switched architecture is that it reduces the amount of transmission an switching to one full duplex channel per site, roughly a factor of three less tha full continuous view. This architecture is better suited to a lecture format than seminar and may necessitate signaling from remote sites to ask a question. offers a degree of interactivity that broadcast alone cannot and is the only prac tical way to offer simultaneous interactive instruction to a large number of site say, anything over five. Although session-switched instruction can run over continuous-view network, the opposite is not the case. If full continuous view meets the design objectives of the IDL project, then the appropriate amount c transmission and switching must be part of the network designed from the star

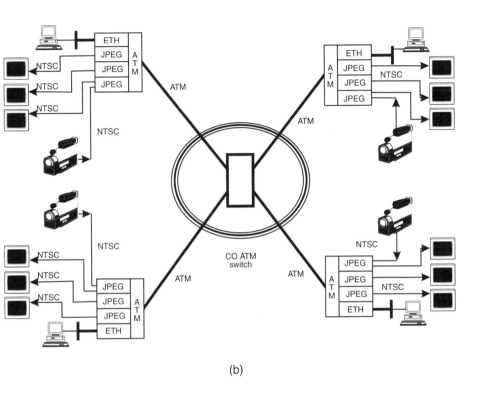

(b)

Figure 10.3 (continued)

10.5.2 Full Continuous View Versus Quad Split

There is a version of continuous view, known as quad split, that can run over one full duplex channel. In that scenario, the video and audio from each of up to four participating sites is brought to a central location and decoded back to analog. From there, each video channel is fed into a four-way quad split processor that digitally reduces the image resolution to 25% of the original. The resulting images are combined into four quadrants and then processed back to analog again. The quad-split analog signal is fed to a digital video codec, which digitizes the signal and feeds it to the appropriate switching equipment for broadcast out to the four participating sites. Audio signals are decoded, bridged, and recombined with the outbound quad-split video in the codecs. Figure 10.4 illustrates a layout of a quad-split network.

Although quad-split processing offers some degree of full continuous view, it takes a toll on image quality and adds to system complexity. There are no fewer than six analog-to-digital/digital-to-analog conversions along with a

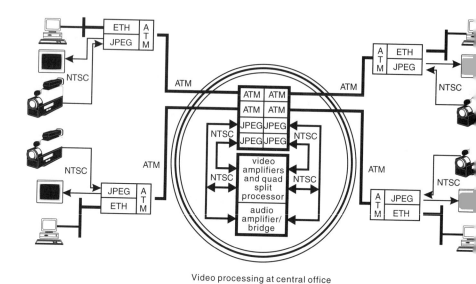

Video processing at central office

Figure 10.4 Quad-split network in detail.

75% reduction in resolution. Depending on the video compression algoritl chosen, the degree of processing can have a major impact on overall pictu quality as well as signal latency. There is the further problem of displaying t quad-split image on a large enough monitor so that all in the class can see. D playing detailed graphics may also be problematic because of the reduction resolution. In that case, the teacher may bypass the quad-split processing a broadcast the video at full resolution to all sites. That is done at the expense interactivity among the participating sites, although audio can remain bridgee

In certain applications, quad-split processing represents a good tradec between interactivity, continuous view, and reduced recurring costs of netwo bandwidth. However, as a general-purpose IDL platform for K–12, it general has not been adopted.

10.5.3 Session Management

The primary function of session management is to provide a user-friend scheduling facility for setting up IDL sessions ahead of time. Session manag ment is essentially a database function that reserves transmission facilities an switch ports to ensure that no blocking occurs at the time of the session. Th function is important for blocking as well as nonblocking networks, since fe things will kill a distance learning program faster than students missing cla because of an error in session management. As a result, session manageme software should be easy to use, preferably menu driven, highly reliable, an ble to schedule a semester's worth of classes ahead of time. It should be able to ring a session on line between several remote sites automatically without involving the instructor in any way. That way, teachers who are not comfortable with the technology can walk into a local IDL classroom with up to three remote sites on line and ready for instruction. Five minutes before the session is o commence, the session management software should signal the instructor.

Session management software should also be capable of handling near and far-end room controls, such as camera positioning, and local and remote video selection of video feeds. For example, a teacher who wants to review the written work of students at a remote site should have the capability of selecting he overhead graphics camera as the remote video feed from that site. When hat activity is done, the teacher can switch back to the room camera. A picture-n-picture option may also be considered useful in certain applications.

Although not all IDL systems will adopt this level of functionality, it will oe important to some applications. It should be clear that session management nvolves far more than just setting up a video teleconference on the fly.

10.5.4 Network Control

Network control must respond to the IDL class sessions stored in the session database. At the appointed time, the appropriate connections are set up between the participating sites of a given IDL session. Network control is generally the domain of the IDL service provider and should be transparent to the end user. It generally involves arcane commands that set up digital cross-connect systems, multipoint control units, or PVCs or SVCs across an ATM fabric.

For a given network architecture, it is possible that circuits may cross several switches before reaching the hub of the instructor's site. As networks grow larger, utilizing more switches to interconnect local IDL hubs, network control and management become increasingly important.

10.6 CLIENT/SERVER-BASED IDL

It is clear to anyone who has spent time with children in front of high-quality CD-ROM source material that the degree of interactivity can be highly engaging and instructional. In that regard, H. Marshall McLuhan once said, "Anyone who believes there is a difference between education and entertainment doesn't know anything about either." However CD-ROM source material in the classroom does have its shortcomings. As mentioned, there is the problem of managing and distributing source material.

The use of computers in classrooms has grown steadily over the last 10 years, with student-to-computer ratios coming down from 125:1 in 1983 to 8:1 in 1994. Schools added over one million computers to classrooms in 1994 alone. The documented impact of computer-based learning in K–12, however has been uneven at best. Many computers in schools today are of older, lower performing vintages. Computers were often introduced into the classroom with out adequate teacher training or compelling applications. Very few are networked, and they often lack such basic features as hard drives, modems, and access to phone lines, much less video capture cards and MPEG decoders. That makes adding networked digital media functionality even more of a challenge.

It would be a reasonable first step to provide a LAN for those computers currently in schools. That action alone moves host-based learning from a self contained, unconnected experience to one that has the potential of creating a new collaborative environment between students and teachers. The value of such an undertaking goes up with the number of computer users, both student and teachers, as well as the degree to which the users are networked to the outside world. The impact of networking among students and teachers in K–12 can be profound, as it has been in the academic and business worlds. The discussion that follows is limited to the implementation of networks that can support digital media.

10.6.1 Digital Media Networks in the School

For reasons already mentioned, client/server–based IDL applications usually require a system design from the ground up, including the host computer, the video server, the LAN, and possibly the WAN. The challenge of doing so should not be underestimated. However, once implemented, a server-based IDL platform has the potential of offering a truly unique on-demand interactive learning experience that can be viewed as an adjunct to the text and e-mail based collaborative work space offered by more traditional data networks. When implementing networked digital media in schools, it should be remembered that this is a fairly leading-edge application, and detailed planning will be critical for success. Working with reputable suppliers and systems integrators is also key.

10.6.2 The Host

Start with an inventory of computing resources on hand, including PCs, workstations, file servers, hubs, WAN access devices, and so on. Anything short of an i486 CPU operating at 25 MHz with 8 MB of RAM should be considered for upgrade. The same is true for an Apple Macintosh that is less than a 68040 at 25 MHz. All hosts will have to have video display capability, lots of memory, NICs, and other associated hardware and software to handle multimedia.

Quality decisions come into play from the start. There are software-only video decode options capable of producing quarter-screen images (240 lines by 360 pixels) and smaller at 8 to 15 frames per second, depending on the host's configuration. That represents a data transfer rate of approximately 100 to 200 Kbps. Improvements in frame rate, resolution, and color depth (bits per pixel) usually require that decompression in hardware to be added to the host. Fortunately, improvements in CPU and bus architectures, along with further reductions in the cost of video decompression hardware, are all favoring better video-handling capability in the host.

10.6.3 The Server

The wide bandwidth and streaming nature of video stresses every aspect of digital media delivery. Starting with disk drives, standard *small-computer serial interfaces* (SCSIs) are capable of delivering roughly 3 MBps of *input/output* (I/O). However, limits in DOS and MacOS operating systems currently throttle sustainable throughput to as low as 400 KBps. SCSI drives on Unix platforms can support a throughput of between 1.5 MBps and 3 MBps, depending on configuration. Standard sequential drives may need to be replaced by a multidrive array to support a large number of multiple streams. Known as *redundant array of inexpensive disks* (RAID), this disk architecture supports data striping over multiple drives and can be optimized for video applications. Improvements in drives and the SCSI bus such as Fast and Wide SCSI-2 should quadruple I/O performance to a theoretical limit of 20 MBps.

The next obstacle in the server is the system bus. Given that the video stream may have to cross the bus several times as it is processed, the actual throughput capability must be several times that of the stream itself. When the support of multiple streams is involved, as in a server, performance of the system bus becomes crucial. Sixteen-bit ISA bus at 5 MBps runs out of gas very quickly. Thirty-two–bit *extended industry standard architecture* (EISA) and PCI local bus, which both support more efficient data handling known as bus mastering, offer throughput in the range of 30 MBps. A Unix-based SPARC-station with S-bus should support a throughput of roughly 30 MBps, depending on the configuration. This kind of performance is necessary to support between 10 and 20 simultaneous MPEG-1 video streams.

Client and server software that optimizes memory management, network access, video streaming, and bandwidth allocation is required to ensure adequate performance over a network. Starlight Networks, IBM, and others offer software that handles networking of digital video streams.

10.6.4 The Network

Nothing will stress a LAN more than a video distribution application that has not been carefully planned for. The bandwidth requirements and streaming nature of digital video present a significant challenge to shared media LANs such as Ethernet. At 10 Mbps (1.25 MBps), Ethernet can theoretically support three to four streams of video. However, because of the way shared media handle congestion and collisions, the practical limit is two streams before audio drop-outs, loss of video frames, and loss of synchronization become a significant impairment. There are ways to improve on this by prioritizing video streams, but overall performance is still fairly limited over standard Ethernet. Furthermore NICs in the server are limited to 1.25 MBps throughput.

To improve throughput and make the most of prioritizing traffic, a high-performance Ethernet switching hub should be considered, especially as the numbers of simultaneous users goes up. By limiting Ethernet segments to one host each, a switching hub offers the necessary performance to support 20 video users. Furthermore, a switching hub can accommodate more than one NIC from the server, which again increases overall throughput for on-demand video-based learning.

When large networks that can support the general distribution of multimedia and video to hundreds of users are being built, ATM LAN backbones will become a necessity. Aside from providing a connection-oriented architecture that can scale to tens of gigabits per second in bandwidth, ATM also offers low latency, and predictable quality of service options that are essential to the streaming nature of video. ATM network interface cards in video servers can now run at a theoretical limit of 155 Mbps (19 MBps). Furthermore, a high-speed, connection-oriented infrastructure such as ATM should go a long way to providing enhanced network management and features such as virtual subnets and tight LAN-WAN integration. As ATM technology matures, it may make sense to consider replacing Ethernet NICs in video-enabled hosts with ATM NICs. The 25-Mbps ATM standard, which can run over category 3 wire, looks like a good candidate for directly attaching ATM to the desktop.

10.6.5 Wide-Area Access

Extending access of stored video material across the wide area is the next logical step. Unfortunately, aside from entertainment-based video on demand, there has not yet been a great deal of work done in this area. One notable exception is the multicast backbone of the Internet. The "M-bone," as it is known among its users, offers fairly low-quality video across portions of the Internet based on IP-multicast. Supporting 3 to 10 frames per second of video at low resolution, it is generally used as a live distribution medium for a community of Internet users running software codecs on high-performance workstations. Access to M-bone video as source material may indeed be desirable; however, given its low overall performance and sophisticated end users, it is not ready as a multimedia distribution model for host-based IDL in K–12.

There are several ways to approach implementing wide-area distribution of client/server IDL and several issues that must be considered. Handling the bandwidth, offering quality of service options, and providing low latency are all key in supporting host-based video applications over the wide area. IP-multicast, for example, is important to many multimedia applications running efficiently over the wide area, and vendors are beginning to include that functionality in their routers. There is growing support for other multimedia-oriented protocols (e.g., PACE, RSVP, SMRP, and ST-2), given the importance this application is taking on in enterprise networks. However, the streaming protocols found in most video servers today do not route over the wide area, and bridging may prove problematic.

If the implementor looks beyond the current model of routed internetworks, integrating ATM in the LAN and the WAN is the long-term solution for distributed client/server IDL applications. Of course, the cost of bandwidth in the WAN is still a big obstacle. However, in the design of private networks, in which the implementor has more of a say in WAN deployment, ATM is capable of offering enough bandwidth to remove the whole notion of distance across the wide area. Video and multimedia servers can now be centralized for ease of maintenance and management of source material. Live video can be distributed to all hosts on the network in real time. And depending on how tightly the ATM LAN and WAN are coupled, the entire network can be managed under one network management system. That will prove key to the overall success of client/server IDL, given the leading-edge nature of these applications.

10.7 PULLING IT ALL TOGETHER

The notion of a detailed design and carefully planned implementation cannot be overstated in delivering IDL networks to video-expert, computer-savvy, K–12 students. Because the value of these systems is in the interactivity they offer, anything that impairs that characteristic will rapidly diminish the return on what will undoubtedly be a substantial investment. In most cases, it would be better to limit the deployment of IDL networks to only the number of nodes that can be "done right." Given the rapid pace of technological change, a phased approach to all aspects of live and client/server–based IDL is all but a given. Testbeds, proof of concept, and pilot projects should be considered necessary stages of any large, long-range project. There are dozens of components that must to be optimized, from disk drives and operating systems to video codecs and LAN/WAN infrastructure.

A client/server IDL network will be only as good as the source material it offers its users, students and teachers alike. Thus, content development will be crucial to the success of those networks. Teacher training is another critical element of IDL for both the live and the host-based applications. If teachers are not comfortable with the tools, their utilization will whither. Having students maintain at least the local part of the network is an idea worth considering. This not only offloads some of the cost of running an IDL network, it offers students valuable technical experience. The result is a sense of ownership of the network by the students and an extended learning community beyond the walls of the local school.

Numerous approaches to financing IDL have included outright state and local ownership of networks; regulatory relief to phone companies who provide the networks; state-mandated tariffed IDL services; block grants from NTIA along with the required matching funds; and regional networks sponsored by state and community colleges. Many RBOCs and independent phone companies will consider offering IDL as part of infrastructure upgrades. They often request regulatory relief in return. Cable TV operators have a long history of offering IDL services; however, those services are often run over analog facilities and thus are limited to a local community. Also, running high-performance internetworks over cable TV facilities may prove problematic. The trend toward deregulation of local and long distance telephone service will no doubt contribute to the number of options available to implementors of IDL networks. And finally, when the design of an IDL network is being contemplated, consideration should be given to how it might be used after hours to defray some or even all of the costs of operation.

Given the impact of video teleconferencing and client/server networks on competitiveness in business today, the argument for making use of those technologies in K–12 IDL networks is compelling. The exponential growth in information and the resulting changes in required skill sets make online access to specialized educational resources critical to the future success of any student today. Furthermore IDL technology has the power to cut across the numerous social and economic barriers to education that many students face, whether they are subjected to the challenges of the inner city or to the isolation of rural environs. It should be clear that IDL has the potential to transform the way students (and teachers) learn and that we are just beginning to utilize the capabilities of this technology.

Case Study of Distance Learning at New York University: The Virtual College

11.1 THE VIRTUAL COLLEGE MODEL

The spread of computer systems and the rise of global markets have rendered traditional bureaucracies increasingly unwieldy. Rigid organizational hierarchies are being replaced by project-oriented work groups that are formed to solve business problems or develop new products and then disbanded when the job is completed. Many participants in these work groups are geographically dispersed across a country or an ocean but are able to collaborate in "virtual" workplaces created by networked computers.

As the physical infrastructure of international business is changing from concrete and steel to computers and communications, teleprograms give practitioners those collaborative and technical skills necessary for working within (as well as on) today's decentralized and networked workplaces, in effect, a virtual college preparing employees for tomorrow's virtual organizations.

This chapter summarizes the activities of The Virtual College teleprogram at New York University during its first two years of development, from the spring of 1992 through the spring of 1994. Much of the work described here was supported by generous grants from the Alfred P. Sloan Foundation.

This chapter was contributed in its entirety by Richard Vigilante, Director of the Information Technologies Institute at New York University's School of Continuing Education. He developed the Virtual College Teleprogram to deliver interactive multimedia telecourses in information technology to student PCs over dialup digital telephone lines. The chapter was included in this book to provide an example of an actual distance learning program.

11.1.1 Telelearning in the Home

Currently, 6 million Americans are enrolled part time in colleges and universities. Over 80% of these part-time students are adults aged 25 and over, and 58% are women. These large numbers mask an even larger adult population that would like to attend college but cannot. Some of the major obstacles faced by adults trying to obtain postsecondary education include:

- Inconvenient class hours;
- Home and job responsibilities;
- Business travel;
- Campus inaccessibility;
- Child or elderly care;
- Physical handicaps;
- Commuting costs.

Communications technologies have been used to deliver higher education to distant learners since the 1920s, when university-owned radio stations first began operation. Successive technologies such as television, time-share computing, and videoconferencing have been utilized to extend the reach of on-campus instruction. Following New York University's introduction of *Sunrise Semester* in the 1950s, televised courses have been the primary means of delivering college instruction into the home. But broadcast television is a largely passive medium, and student interactions with faculty and other students are limited to the occasional phone call or letter. Videoconferencing systems have made televised instruction interactive, but only those students who work at or can get to a business or university videoconferencing site can participate in the courses.

The introduction of *computer-mediated communications* (CMC) systems such as computer conferencing and e-mail in the 1980s finally permitted the delivery of interactive instruction directly into the home. Colleges such as the New Jersey Institute of Technology, the New School for Social Research, and the University of Phoenix have used CMC packages like EIES, Unison, and Caucus to offer online courses and some degrees over the public telephone network to students with home PCs and modems.

11.1.2 Limitations of Current CMC Software Packages

While the current generation of computer-mediated communications packages support a wide range of student-faculty interactions, they have the following important limitations as distance education delivery systems:

- *Limited capabilities.* Current CMC packages are essentially text-based communications systems with limited graphics and database capabilities. Most of today's innovative on-campus computer-based training programs utilize multimedia and hypertext systems to enhance learning.
- *Hybrid delivery.* Current CMC packages cannot provide the myriad educational support services (e.g., student recordkeeping, curriculum development, textbook delivery) characteristic of all college operations and therefore require mail, express, or fax delivery of instructional and administrative print materials to and from students and faculty.
- *Discussion focus.* Current CMC packages allow students and faculty to *discuss* class projects such as information systems, engineering designs, and marketing presentations but do not allow them subsequently to *develop* those projects. The electronic equivalents of such on-campus facilities as laboratories, workshops, and studios are missing.

11.1.3 The Virtual College and Lotus Notes

In an effort to meet the needs of a broad range of distant learners and to deliver high-quality, interactive instruction directly into the student's home, New York University's School of Continuing Education began work in 1991 on The Virtual College teleprogram. The Virtual College was intended to be a comprehensive instructional management system for the efficient production and delivery of a potentially wide range of college, business, and technical courses and programs. Through The Virtual College, students would receive instruction, ask questions, conduct analyses, resolve problems, and complete projects—all largely at their own convenience and from practically anywhere in the world.

In addition to being a higher-education delivery system, The Virtual College was intended to be a model for corporate training. Corporate training and professional education is a $100-billion-a-year business in the United States. Upward of 35 million individuals receive formal, employer-sponsored education each year. The facilities, staff, and travel costs of traditional training programs are considerable. To meet the continuing demand for training, education has to be constantly available to employees through convenient and economical means. The cost effectiveness of online versus on-site training is of increasing interest to thousands of business and public organizations.

Given the limitations of existing CMC software packages, the Lotus Notes groupware package was selected to be The Virtual College's software platform. First introduced in 1989, Notes is a powerful group information manager that gives people who work together an electronic environment in which to collect, organize, and share information, using networked PCs. For The Virtual College, Notes provided the following advantages over current CMC packages:

- *Group communications.* While providing computer conferencing and e mail functions similar to most CMC software, Notes interactions incorpo rate data, graphics, audio, and video in addition to plain text.
- *Database management.* Notes supports the development, sharing, and up dating of large multimedia databases to distribute information sets among all users. Notes databases incorporate a broad range of hypertext, search and report functions.
- *Applications development.* Notes provides a robust but user-friendly capability to build working applications. As course projects, these ap plications could include information systems, marketing presentations concurrent engineering efforts, and environmental analyses. As adminis trative tools, these applications could include student recordkeeping, cur riculum development, and library services.

11.1.4 Initial Virtual College Telecourses

The Virtual College is being developed as the electronic equivalent of a tradi tional college. As shown in Figure 11.1, The Virtual College system mode consists of four Notes database modules: Course, Faculty, Library, and Admini stration. All the services indicated will be provided on line using Lotus Notes to participating students, faculty, and administrators over national and interna tional public data networks. Those networks, and the attached Notes servers and user PCs, would constitute The Virtual College's campus.

- *Course databases* are the Virtual College's focal point and support all stu dent-faculty discussions, case study analyses, and project developmen for each telecourse. Each course database has an outline, session guides and Notes forms for discussions, analyses, exams, and evaluations. A separate database called *The Virtual Cafe* provides for informal and extra curricular student-faculty discussions.
- The *Faculty database* supports all faculty discussions and development relating to curriculum, student, grade reporting, and policy matters.
- The *Library database* contains all licensed electronic copies of published books and articles, telecourse hypertexts and case studies, and digital audio and video materials. The database also provides an online gateway to external electronic libraries and databases.
- The *Administration database* supports student admissions, registration and related recordkeeping and serves as an interface to official university student information systems.

The Virtual College opened its electronic doors in the spring 1992 semes ter. The teleprogram utilized an IBM PS/2 model 95 PC as a server and pro vided student access over a three-line, national 800-number communication.

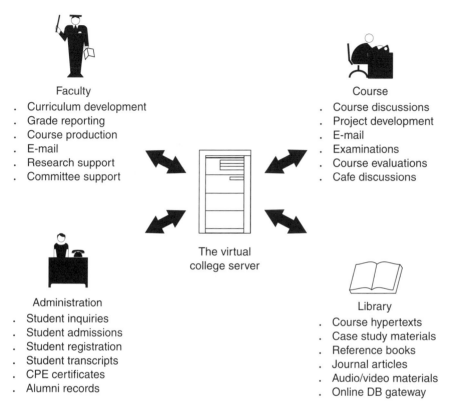

Faculty
- Curriculum development
- Grade reporting
- Course production
- E-mail
- Research support
- Committee support

Course
- Course discussions
- Project development
- E-mail
- Examinations
- Course evaluations
- Cafe discussions

The virtual
college server

Administration
- Student inquiries
- Student admissions
- Student registration
- Student transcripts
- CPE certificates
- Alumni records

Library
- Course hypertexts
- Case study materials
- Reference books
- Journal articles
- Audio/video materials
- Online DB gateway

Figure 11.1   The Virtual College delivery system.

network. The initial noncredit telecourse enrolled 10 students and was an introduction to information systems analysis and design. During the eight-week telecourse, the students and the instructor formed a virtual workgroup and collaborated on the development of a data collection system in Notes.

Beginning with the fall 1992 semester, a new four-credit graduate course was offered to provide eight students with a working knowledge of the techniques and technologies used to develop virtual workplaces that connect people as well as computers. Students and the instructor collaborated online to analyze, design, and build "corporate cyberspaces" for both their own and case study organizations. Six students enrolled in this graduate course during the spring 1993 semester. Figure 11.2 shows a sample Notes screen from the graduate telecourse that incorporates text and a still video image.

The initial group of 24 Virtual College students were primarily midcareer professionals and managers from such large organizations as NBC, ITT, Con Edison, the United Nations, ChemBank, and British Airways. Most were generalists and had limited experience with networked computing. Their primary

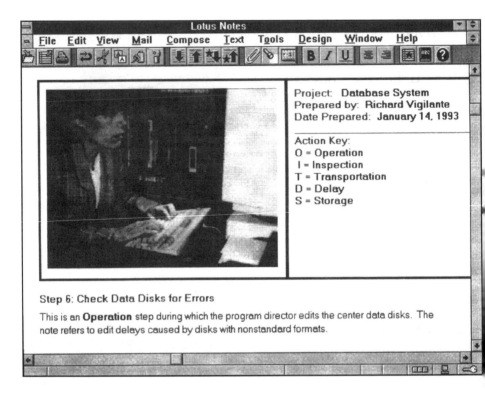

Figure 11.2   Sample Notes screen from Virtual College telecourse.

reason for enrolling was a professional curiosity about the process and the potential of virtual workgroups.

11.2   DEVELOPMENT OF A GRADUATE TELEPROGRAM

11.2.1   Faculty Training: On Site Versus Online

Following successful delivery of the initial pilot telecourses, the next logical curricular step for The Virtual College was the development of a complete multicourse, multifaculty graduate program. Funded by a $27,300 grant from the Alfred P. Sloan Foundation, the project used NYU's existing, on-campus *Advanced Professional Certificate* (APC) in Information Systems Auditing as the basis for the new Virtual College teleprogram. Since its inception in the summer of 1990, the 16-credit, four-course Systems Auditing APC enrolled over 120 graduate students.

In June 1993, a display advertisement for part-time teleprogram faculty appeared in *The Chronicle of Higher Education*. Over 30 resumes were received and from those, the following three individuals were selected to join the principal investigator in becoming the first Virtual College faculty. Because distance education pertains to faculty as well as students, the new instructors had a diverse geographic background.

- Professor of computer science at San Francisco State University;
- Education director of a Washington, D.C., professional association;
- President of a New Hampshire systems consulting firm.

The original plan was to have the new faculty meet with the teleprogram director in New York for a multiday Lotus Notes training and curriculum planning session. The demonstrated ability of students in the initial pilot telecourses to learn Notes on line as part of the course activities led to a modification in the faculty training plan. Two instructors would receive an on-site two-day Notes training session. The third instructor, however, would learn Notes on line by being an active participant in the fall 1993 telecourse. All instructors were to receive training equivalent to the *Lotus Notes End-User Course* prepared by Lotus Development Corp.

11.2.1.1 On-Site Faculty Training in Notes

Two instructors received an individualized, 12-hour, hands-on training session using materials and exercises contained in the *Lotus Notes End-User Course* package. The session gave the instructors a sufficient technical knowledge of Notes communications, functions, and applications development to subsequently develop their own Notes-based telecourses. The topics covered in the session are listed below:

Lesson 1: Getting Started (1 hour)
- Getting Started in the Notes Workspace
- Using the Mouse to Perform Basic Operations
- Working in the Notes Workspace
- Working with Windows

Lesson 2: Understanding Documents and Databases (1 hour)
- Adding and Opening a Database
- Opening Documents in a Database
- Using the Help System
- Viewing a Database from Different Perspectives
- Working with Multiple Windows

Lesson 3: Composing and Editing Documents (1 hour)
- Composing a Main Topic Document
- Composing and Formatting a Response Document
- Editing Documents in a Customized Database

Lesson 4: Advanced Editing and Printing (1.5 hours)
- Exploring Word Processing Tools
- Printing and Setting Print Formats

Lesson 5: Importing and Exporting Files; Creating DocLinks (1.5 hours)
- Importing and Exporting Files
- Using and Creating DocLinks

Lesson 6: Managing Your Workspace; Creating a Private View (1.5 hours)
- Customizing the Workspace
- Managing Databases in the Workspace
- Creating a Private View
- Controlling the System Environment

Lesson 7: Using Notes Mail (1 hour)
- Exploring a Mail Database
- Composing and Sending Mail
- Managing Mail

Lesson 8: Creating a Notes Database (3.5 hours)
- Planning a Notes Application
- Creating a Database from a Template
- Creating a New Database
- Creating a Notes Form
- Creating a Notes View
- Completing the Database

11.2.1.2   Online Faculty Training in Notes

The third instructor was to be trained in Notes as all students were: by actively participating in an ongoing telecourse. The advantage of this method (if suc cessful) was that new faculty could be trained in telecourse tools and design from their home PCs without the need to travel to a remote training site. This training mode required faculty, as with students, to assume a "role" in the tele course—a rationale to engage them in telecourse interactions from the begin ning and to use those interactions as a basis for learning Notes. For telecourse students, their roles as members of project workgroups had proved to be a successful way to learn Notes from the first spring 1992 telecourse.

For the instructor in the fall 1993 telecourse in Systems Analysis, the role was that of *client*, the person for whom the case study systems project was being developed by the students. The client role required a good familiarity with the case study's operational parameters and problems. Close and ongoing communication via e-mail between the course instructor and the client–trainee instructor was also required to provide additional case study information and formulate appropriate responses to student inquiries. Beyond that, the client–trainee instructor needed no special Notes training other than that being given to students as the telecourse progressed. Figure 11.3 shows a Notes screen image of a typical response to students from the instructor-as-client.

This method of providing Notes training to faculty was highly successful. By his ongoing participation in the six-week telecourse, the instructor received far more hands-on experience with Notes than did his colleagues in the two-day training session. In addition, the online trainee received an intensive les-

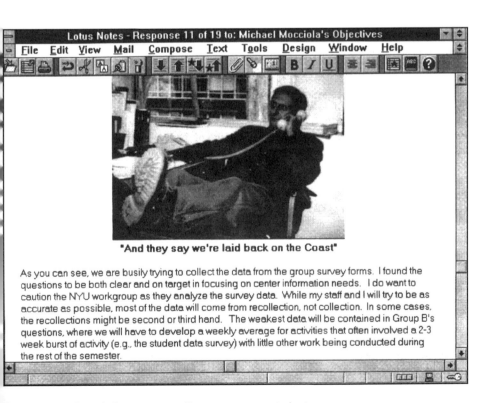

Figure 11.3 Sample instructor-as-client response to student.

son in Notes-based telecourse design and delivery issues that were usefu guides in subsequently developing his own telecourse.

As a result of this training experience, The Virtual College will require al new faculty to participate in some capacity in an ongoing telecourse before in dependently teaching their own telecourse. While the role of client was apprc priate for the Systems Analysis telecourse, other roles may be required fo faculty trainees participating in different courses. Among the roles that ma prove equally effective in other course environments are laboratory instructo coteacher, guest lecturer, and group tutor.

11.2.2 Developing the Teleprogram Curriculum

In 1990, New York University's School of Continuing Education began offerin a four-course, 16-credit APC in Information Systems Auditing. This new gradu ate program was designed to prepare auditors and analysts for information sys tems audit and control responsibilities. In the initial design of the program little emphasis was placed on familiarizing students with broader managemen information systems concepts and methodologies.

In recent years, auditors at all levels and in all functional areas increas ingly have had to work with and on complex computer and communication systems. This changing responsibility is reflected in the certification require ments of the *Institute of Internal Auditors* (IIA). The IIA is the key professiona association in the audit field, and it administers the *certified internal audito* (CIA) examination to certify the professional competence of auditors in the fo lowing disciplines:

- Internal auditing administration;
- Electronic data processing (EDP) auditing and statistical sampling;
- Organizational behavior and management;
- Management information systems;
- Economics and finance;
- Accounting.

The CIA exam consists of four equal parts covering these disciplines. ED auditing topics have constituted 10% to 20% of the Part II questions in recen CIA exams. Management information systems topics have constituted 40% t 50% of the Part III questions in recent CIA exams. The general topic areas fo EDP auditing and management information systems are indicated below.

EDP Auditing (10%–20% of CIA Part II Questions)
- Effect of EDP on Controls and Audit
- Nature and Audit of Computer-Related Controls
  - General Controls

- Organization and Operation Controls
- System Development, Changes, and Documentation Controls
- Hardware and Systems Software Controls
- Security and Access Controls to Computer Facilities, Programs, and Data
- Data and Procedural Controls
  - Application Controls
    - Input Controls
    - Processing Controls
    - Output Controls
    - File and File Management Controls
- Audit of Computer Processing
  - Auditing Computer Programs
  - Auditing Computer Files
  - Auditing Computer Processing Systems
- Control and Audit of Advanced Systems
- Control and Audit of Third-Party Systems
- Control and Audit of Mini- and Micro-computers
- Feasibility Studies and System Conversion
- Evaluating System Efficiency and Effectiveness
- Computer Abuse and Crime
- Establishing and Maintaining an EDP Audit Function

Management Information Systems (40%–50% of CIA Part III Questions)

- System Fundamentals
  - Information Systems Concepts and Techniques
  - The Systems Concept and Systems Theory
  - Management Decisions and Information
  - Information Systems Environment and Changes Therein
  - Elements and Operations of Management Information Systems
  - Behavioral Considerations
- Business Computer Systems and Their Components
  - Forms and Operations of Computer Hardware
  - Computer Languages and Programs
  - Forms of Input, Processing, and Output
  - Storage Concepts and Data Files
  - Computer Operations and Procedures
  - Operating Systems
- File Management and Data Base/Data Communication Systems
  - File Organization and Data Management Systems
  - Major Supporting Software Systems
  - Data Base/Data Communication Systems

- Systems Analysis and Design
  - Systems Development Life Cycle
  - Systems Analysis and Selection
  - Systems Implementation and Evaluation
- Various Information Systems in Use

The faculty found that the original APC course sequence provided gradu ate audit students with insufficient study of the management information sys tems area and offered excessive and redundant study of the EDP auditing area During the fall 1993 semester, the four teleprogram faculty used The Virtua College's Faculty Database to review and revise the APC course scope and se quence on line. In addition to the CIA examination requirements outlinec above, the faculty analyzed the following three model systems curricula devel oped by professional associations:

- *Model Curriculum for Information Systems Auditing* (developed by the In stitute of Internal Auditors);
- *Graduate Curriculum in Information Systems* (developed by the Associa tion for Computing Machinery);
- *Model Curriculum for Graduate Computer Information Systems* (devel oped by the Data Processing Management Association).

Print copies of these documents were distributed to all faculty in Septem ber 1993. During a three-month curriculum development period, the Notes Fac ulty Database was used by all four instructors in an online collaboration tha identified, analyzed, and selected the appropriate course topics, assignments readings, and case studies for the new APC teleprogram. During that period, the faculty used the ongoing fall 1993 Systems Analysis telecourse as a real-time laboratory to develop a uniform telecourse curriculum structure and delivery model. Each new APC telecourse would incorporate a course outline, six course-session guides, and a detailed case study.

During their curriculum development effort, the four instructors made ex tensive revisions to the APC program. The revised APC course sequence wa: designed to provide audit students with a balanced study (as identified by IIA certification requirements) of both the management information systems anc EDP auditing areas. The four new APC telecourses are described below.

- Systems Analysis (Y52.1100/4 credits). This course is an introduction tc the role of information technology in organizations and an overview of the systems development life cycle. Topics include planning and managing the information resource; identification of user information requirements evaluating the costs and benefits of systems alternatives; application o structured systems analysis tools and processes; input and output design considerations; and building and testing prototype systems.

- Database Management (Y52.1110/4 credits). This course provides a working knowledge of the logical structure and physical implementation of database management systems. Topics include information as an organizational resource; concepts of data structures; the application of data models and dictionaries in database design; strategic data and systems planning process; data normalization steps; data administration; and client/server and distributed database systems.

- Systems Auditing (Y52.1130/4 credits). This course focuses on the standards and techniques employed in the audit and control of information systems. Topics include the nature and audit of computer-related controls; general controls (hardware, software, security, and access); application controls (input, database, processing, and output); control and audit of end-user, networked and client/server systems; and evaluation of overall system efficiency and effectiveness.

- Information Security (Y52.1140/4 credits). This course is an introduction to the technologies and management of business telecommunications networks and the security issues surrounding their use. Topics include network technologies and topologies; communications media and carriers; the Open System Interconnection (OSI) model; physical and logical data and systems security; and contingency planning processes and strategies.

The instructors believed the revised APC program provided a balanced curriculum of both management information systems and EDP auditing courses. The original APC subject title "Information Systems Auditing" was felt to be too narrowly focused on the EDP auditing field and not properly reflecting the inclusion of more generic information systems courses. To more accurately reflect the proposed information systems/EDP auditing curriculum, the subject title for the revised APC was changed from "Information Systems Auditing" to "Information Technology." Information technology is a widely accepted designation in the audit profession that encompasses both the management information systems and the EDP auditing areas.

11.2.3 Anatomy of a Telecourse

During the fall 1993 semester, the prototype APC telecourse was offered to 21 students. This graduate course in Systems Analysis served as both a faculty training and a curriculum development platform for the new faculty's design of the APC teleprogram. In terms of course structure, full enrollment, participant collaboration, communications costs, and student evaluations, this Systems Analysis course embodied and tested all the operational attributes of the new APC teleprogram in Information Technology.

Table 11.1 shows the job titles and employers of the 21 fall 1993 tele-course students. All students were employed full time, and all had at least a bachelor's degree (about 20% had a master's degree, typically an MBA). Five of the students were "sponsored," that is, their companies paid the $1,900 tele-course tuition upfront (at registration). Most of the others had their tuition paid by their companies' tuition reimbursement programs after final grades were submitted to their personnel departments. All the students were PC literate (i.e., in word processing and spreadsheets), but only six of them had some professional systems experience—the rest were nontechnical managers or professionals.

Table 11.1
Job Titles and Employers of Fall 1993 Telecourse Students

Programmer/analyst	Time, Inc.
Accountant	JMW Consultants, Inc.
Staff manager	NYNEX TeleCommunications Group
Trainer/consultant	Self-employed
Program manager	DSTI
Technical trainer	Chemical Bank
Staff manager	NYNEX TeleCommunications Group
Vice president	Bankers Trust Co.
Director of Financial Planning	Russell Reynolds Associates
Staff manager	NYNEX TeleCommunications Group
Project manager	Marsh & McLennan, Inc.
Assistant vice president	Marsh & McLennan, Inc.
Technology consultant	Moody's Investors Service
Account executive, ISDN	NYNEX
Analyst	Morgan Stanley & Co.
President	Crown Properties
Network administrator	Memorial Sloan-Kettering Cancer Center
Manufacturing supervisor	R. R. Donnelly & Sons Co.
Software development manager	Moody's Investors Service
System administrator	NYNEX Telesector Resources Group
Systems manager	Moody's Investors Service

During the six-week core of the telecourse, students, the instructor, and the "client" collaborated on line to analyze, design, and build a working information system in Lotus Notes. Because of the relatively large class size, the students were divided into four subgroups to work concurrently on various phases of the case study systems project. During a two-week period when the subgroups were conducting the preliminary and alternative analyses phases of the project, 750 Notes documents were created! This was an average of 35 questions, proposals, analyses, resolutions, and so on, per student—a level of participation that would be rare even in the most active on-campus seminars over a similar time period.

A total of 1,843 documents were created by telecourse participants. Table 11.2 shows the number of documents generated in the seven telecourse Notes databases.

Table 11.2
Number of Documents Created by Notes Database

Telecourse Database	Documents
Course Database	324
Workgroup A	497
Workgroup B	137
Workgroup C	163
Workgroup D	286
The Virtual Cafe	256
Mail	180
Total	1,843

Functioning as members of their virtual project teams, the students established discussion guidelines, critiqued and edited each other's work, managed virtual workplace responsibilities, and at times ran an asynchronous groupware package as if it were an online chat service. This level of interaction was maintained even when many students were away on business trips—one as far as Tokyo!

As with previous telecourses, students used an AT&T 800 ReadyLine service that provided them with free and unlimited access to the Virtual College server (over a three-line hunt group with 14,400-baud modems) throughout the 50 states, Puerto Rico, the U.S. Virgin Islands, and Canada. Students traveling outside the United States were asked to use AT&T's USADirect service, which would connect them to the NYU 800 number from over 100 countries. The advantage of USADirect was that students could connect from practically anywhere and could utilize the speed of the 14,400-baud modems that many of their notebook PCs had. It proved to be very reliable. The disadvantage was that international calls were not free (unlike the U.S. 800 service) and averaged $1.00 per minute from Europe. Because most telecourse database replications took only a few minutes at 14,400 baud, the total cost was minor for traveling U.S. students. Costs for resident international teleprogram students, however, would be another matter.

During the overall nine-week duration of the telecourse (October 11–December 12), 2,604 calls totalling 159 hours were made to the Notes server. Of that total, 1,858 calls (71%) were made from the New York metropolitan region, and the remaining 746 calls were made from 18 states, Puerto Rico, Canada, and Japan. Using the AT&T 800 service, the total telecommunications cost was $1,820, or $87 per student. The average call duration was 3.7 minutes, and each student made an average of 100 calls during the course.

Table 11.3 shows the daily distribution of calls during the six-week instructional core (October 11–November 21) of the telecourse. Logon activity clearly peaked on those days (indicated by asterisks) when student workgroup assignments were due. The busiest day was Sunday, October 24, when 185 calls were made by students submitting their preliminary analyses for the case study systems project.

Table 11.3
Distribution of Fall 1993 Telecourse Daily Calls

October	Number of Calls	November	Number of Calls
11 (Monday)	52	1 (Monday)	76
12 (Tuesday)	16	2 (Tuesday)*	170
13 (Wednesday)	31	3 (Wednesday)	68
14 (Thursday)	28	4 (Thursday)	40
15 (Friday)	23	5 (Friday)	31
16 (Saturday)	25	6 (Saturday)	70
17 (Sunday)	45	7 (Sunday)*	133
18 (Monday)	37	8 (Monday)	91
19 (Tuesday)	42	9 (Tuesday)	62
20 (Wednesday)	34	10 (Wednesday)	42

October	Number of Calls	November	Number of Calls
21 (Thursday)*	60	11 (Thursday)	34
22 (Friday)	46	12 (Friday)	38
23 (Saturday)	58	13 (Saturday)*	51
24 (Sunday)*	185	14 (Sunday)	40
25 (Monday)	48	15 (Monday)	37
26 (Tuesday)	61	16 (Tuesday)*	66
27 (Wednesday)	75	17 (Wednesday)	40
28 (Thursday)	61	18 (Thursday)	42
29 (Friday)	54	19 (Friday)	21
30 (Saturday)	81	20 (Saturday)	30
31 (Sunday)	122	21 (Sunday)	51

* Due dates for student workgroup assignments.

11.2.4 Student Evaluations of Fall 1993 Telecourse

Following completion of the fall 1993 Systems Analysis telecourse, the students were asked to complete an online evaluation of the course using a Notes evaluation form. The evaluation consisted of 23 multiple-choice questions and an optional written evaluation. As shown in Figure 11.4, the scale values for each multiple-choice question were described in a Help line at the bottom of the display screen. The questionnaire was completely anonymous, and all completed evaluations were sent to a public display area of the course database.

The 15 questions on course content and instructor performance were similar to those asked of NYU continuing education students in on-campus courses. The telecourse students' evaluations of these factors were quite positive, with the content and performance items averaging an 85% response of *Very Good* or *Excellent*. The responses to the eight questions that evaluated Lotus Notes as an instructional delivery system were equally favorable. Some of the highlights from that part of the student evaluation follow:

- 93% found Notes somewhat or very easy to use.
- 72% always or most often worked on the course from home.
- 51% occasionally or never printed course documents.
- 86% generally worked on the course from 6 PM to midnight.
- 79% spent seven or more hours per week on the course.
- 57% found the course graphics and imagery generally or very useful.

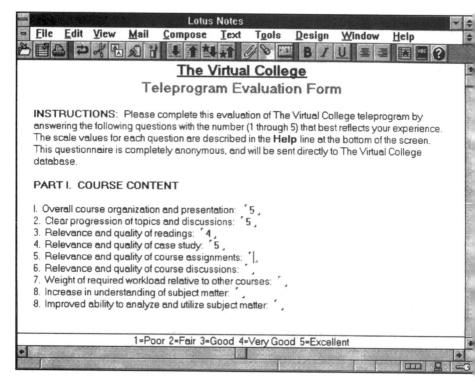

Figure 11.4 Display of Student Evaluation form (cursor at question 5).

Part I Evaluation Summary: Course Content

1. Overall course organization and presentation:
   Poor-0%; Fair-0%; Good-7%; Very Good-57%; Excellent-36%

2. Clear progression of topics and discussions:
   Poor-0%; Fair-0%; Good-21%; Very Good-43%; Excellent-36%

3. Relevance and quality of readings:
   Poor-0%; Fair-0%; Good-14%; Very Good-50%; Excellent-36%

4. Relevance and quality of case study:
   Poor-0%; Fair-0%; Good-0%; Very Good-57%; Excellent-43%

5. Relevance and quality of course assignments:
   Poor-0%; Fair-0%; Good-14%; Very Good-43%; Excellent-43%

6. Relevance and quality of course discussions:
   Poor-0%; Fair-0%; Good-7%; Very Good-43%; Excellent-50%

7. Weight of required workload relative to other courses:
Very Light-0%; Light-0%; Average-50%; Heavy-36%; Very Heavy-14%

8. Increase in understanding of subject matter:
Poor-0%; Fair-0%; Good-7%; Very Good-50%; Excellent-43%

9. Improved ability to analyze and utilize subject matter:
Poor-0%; Fair-0%; Good-14%; Very Good-57%; Excellent-29%

Part II Evaluation Summary: Instructor Performance

10. Clarity and presentation of subject matter and topic discussions:
Poor-0%; Fair-0%; Good-7%; Very Good-50%; Excellent-43%

11. Ability to explain difficult or unclear topics:
Poor-0%; Fair-0%; Good-14%; Very Good-36%; Excellent-50%

12. Knowledge of subject matter:
Poor-0%; Fair-0%; Good-7%; Very Good-36%; Excellent-57%

13. Responsiveness to questions and comments:
Poor-0%; Fair-0%; Good-7%; Very Good-29%; Excellent-64%

14. Clarity of instructor's expectations:
Poor-0%; Fair-0%; Good-28%; Very Good-36%; Excellent-36%

15. Ability to motivate and manage workgroup discussions:
Poor-0%; Fair-0%; Good-14%; Very Good-50%; Excellent-36%

Part III Evaluation Summary: Delivery System

16. Overall evaluation of Notes for distance education:
Poor-0%; Fair-0%; Good-21%; Very Good-36%; Excellent-43%

17. Overall experience in using Notes for course:
Very Difficult-0%; Somewhat Difficult-0%; Average-7%;
Somewhat Easy-64%; Very Easy-29%

18. Communications problems with server (busy signals,
disconnections, etc.):
Very Frequent-0%; Somewhat Frequent-14%; Occasional-21%;
Infrequent-57%; Never-7%

19. How often did you work on course from home:
Always-36%; Most Often-36%; About Half-21%; Occasionally-7%;
Never-0%

20. How often did you print course documents:
Always-14%; Most Often-14%; About Half-21%; Occasionally-37%;
Never-14%

21. At what time of day did you generally work on course:
6 AM to Noon-7%; Noon to 6 PM-0%; 6 PM to Midnight-86%;
Midnight to 6 AM-0%; No Regular Time-7%

22. How much time per week did you spend on ALL course work:
0–2 Hours-0%; 2–5 Hours-0%; 5–7 Hours-21%; 7–10 Hours-29%;
10+ Hours-50%

23. Instructional value of graphics and imagery in documents:
No Value-0%; Little Value-14%; Occasionally Useful-29%;
Generally Useful-14%; Very Useful-43%

Part IV Evaluation Summary: Written Comments

As an optional part of the evaluation, students were asked to write any additional comments they had about the Systems Analysis telecourse. These comments are provided in their entirety below.

Student A

"I *really* enjoyed the experience. At first I was apprehensive not knowing what to expect, but nonetheless was lured by the challenge of something new: working in a virtual environment. Looking back I must say while I felt a bit awkward in this new environment, I feel it is a viable medium in which to work or learn in. Going forward I will be taking more classes with the hope of applying this virtual concept within my business unit. My sense of apprehension now has me totally intrigued."

Student B

"Utterly fascinating and quite valuable. Not only did this course provide a background in the mechanics of telecollaboration, but it also provided very interesting insights into how people work together in this way. The observations that one could draw (and that were drawn in the Sproull & Kiesler book) are many and important. I learned about how people interact in this setting, what people can reasonably accomplish, etc. I still don't believe that you can develop end-user software packages this way (database designs, maybe), but I do believe that this mode of communication is good for many things.

"I'm going to be building various Notes applications during the coming year or two; I can easily learn how to build them on my own, but this course has given me an indication of what to expect from the applications that I could never get from a book or seminar.

"I had two difficulties, however:

1. The pace was quite fast, making it difficult to keep up when you have a demanding, full-time job plus (in my case) a moonlight gig. It's harder to deal with deadlines when everyone is on their own personal schedule.
2. Notes databases have what I'll call a 'lack of linearity'; i.e., it's difficult to tell where to look or what to do 'next.' You end up having to jump around from place to place and often forgetting where you are. DocLinks help, but you're dependent on people remembering to put them in, a task that is unnatural to those without experience. I don't have any brilliant ideas on how to solve this, though things got better as the course went on (you get better at jumping around and remembering context, and you also get better at putting in DocLinks)."

Student C

"This course was instrumental in making me become aware of the potential value in working in a virtual office. Because I worked on the course from my home, I learned that working at home has its plusses and minuses and that I can discipline myself to this environment. Having the Virtual Cafe was also very informative. I looked forward to communicating extensively with people I had never met face to face. I hope that the curriculum will expand to more telecourses.

"The course pace was hectic and for me working full time while going to school, and not being able to access the server from work created a challenge in learning Notes. However, it helped me to organize and discipline myself to complete the assignments. In the beginning I was apprehensive but as the project unfolded, I gained inner confidence. Now, I will do more with Notes at work instead of looking at it as a desktop accessory. Thank you."

Student D

"That first week, I felt like a voyeur. I kept reading all the Responses without being able to express my own thoughts satisfactorily. Working online in workgroups definitely takes some getting used to. However by now I really am quite comfortable with this medium. I find that I am not printing out many documents anymore whereas in the beginning I printed everything. The case study required more of a working knowledge of technology than I was comfortable with. I also felt that not having reference books on Lotus Notes aided considerably in the learning process, I was forced to figure it out one way or another and the online help was great. Overall, I felt this course was well worth the time and effort. I would also be very interested in any other telecourses both for the convenience and the practice in Virtual Communications."

Student E

"Students would have benefited from receiving more Notes documentation."

Student F

"Overall an excellent course, very stimulating."

Student G

"This course certainly helped me in my job. I learned a lot about Notes from a practical standpoint that I would not have learned in taking a 'generic' Lotus Notes course. I would have liked to have a user's guide for Notes. When experiencing software problems, it would be nice to have a number to call for assistance, especially during the evening and/or weekends."

Student H

"Six weeks was too short. Discussions and alternatives on assignments could not be pursued in any length or depth.

"The time requirements of the course should be specified more realistically in the course description. Fortunately, I had the time to spend on the course. Others in my group, apparently, did not.

"The course director should encourage participants to publish their intended log-in schedules as soon as groups are formed. This will give the active participants some idea of how long to wait for responses and new ideas. (Many of the groups seemed to have discovered this on their own, but because the class is so short these types of tips and guidelines should be provided by the instructor.)

"It may be useful to have each group select a leader early on.

"Strict adherence to deadlines was good, but you should make that policy clear up front."

Student I

"I feel that the course helped present a real-life scenario in real-world timeframes. It might have been nice to extend the course by a week or two to allow more discussion on topics but I feel that it would have been unrealistic. The course is very effective in getting together people on a project in a constrained timeframe. The reading material helped put framework and thought provocation to enable synergy to take place.

"I would recommend this course highly to those that are attempting to put together groupware communication strategies, not for the Notes content—which was good—but for the information contained in the reading material. This course could be marketed to the decision-makers in the corporate world who do not have the time to attend regular classes.

"I really enjoyed the course."

Student J

"This course has provided a wonderful opportunity to experience virtual work-group collaboration. I was very impressed with what a group of strangers, mostly unfamiliar with Lotus Notes, was able to accomplish. I enjoyed taking the course and feel that I have made very good friends with people that I might never meet. I'll miss all of you!"

Student K

"I got more from this course than most other courses I have ever taken and look forward to more online courses. My only complaint would be the amount of time that we had to get into the really fine details. It seemed that a lot of information for the case study was missed due to time constraints."

Student L

"Overall, I enjoyed and was very impressed with the telecourse, and in addition to being interested in followup courses on Notes development, I'd also like to see the Notes vehicle used for conducting other, non-systems-related classes. (How about a creative writing class in Notes, for example?)"

Student M

"I could definitely have used more time. The sessions beginning with designing a database and building forms and views were the toughest learning curve for me and merited more time to work through the steps, more mentorship in developing them, and more chances to see what others were achieving and to cross-educate as we had with other steps in the course.

"I concur with whomever made the suggestion of expanding telecourse offerings at NYU to nontechnical disciplines. I would eagerly consider and recommend these.

"Thanks!"

11.3 INTERACTIVE VIDEO TELEPROGRAMS

11.3.1 The Value of Instructional Video

The effectiveness of video as a tool for training has been demonstrated at every level of education, from preschool through adult education. Video shifts time, space, and content into viewer-adjusted, viewer-accepted segments, making information available whenever, wherever, and in whatever amounts needed. It uses color, motion, and sound in one universally acceptable medium to convey technical skills, concepts, and attitudes equally well. Whether broadcast, cablecast, or taped, video brings the advantages of compression, consistency, and visualization to the learning process.

Instructional video offers the advantage of compression, eliminating unnecessary material and delivering only as much information as is directly related to course objectives. With compression, students can learn in one-third to one-fourth the time it would usually take in a classroom setting. Another advantage of instructional video is its ability to deliver information consistently—all students receive the same content in the same style of delivery. Individual instructor differences in the interpretation, style, or emphasis of course content is eliminated.

Video is a visual medium and has the ability to hold student interest through color, motion, and sound. Visual images can add to student retention and recall of information. Color visuals can help illustrate difficult concepts, clarify information, accelerate learning, improve comprehension, increase recognition, and reduce errors. The motion element of video facilitates presentation of subject demonstrations, process flows, and role simulations. Finally, audio can reproduce various voice and sound conditions and can enhance student recognition and discrimination.

11.3.2 Video Delivery to Home-Based Learners

Video telecourses have been broadcast to home-based learners for over forty years. The widespread availability of home videocassette recorders and personal computers has introduced various degrees of student control of and interaction with video courseware. Current consumer technologies support three models of video instruction in the home—Video and Print Telecourse, Video and Conferencing Telecourse, and Interactive Video Telecourse.

11.3.2.1 Video and Print Telecourse

This is the most common model of home-based video instruction. Characterized by the dozens of telecourses funded through the Annenberg/CPB Project, video and print courses consist of 8 to 12 hours of presentation video, one or more textbooks, and a course study guide. Many students use home VCRs to watch the telecourse sessions, which allows them to stop and replay sections of the tapes they don't fully understand. Where concepts are still not understood, however, students must rely on course textbooks, which are often only loosely linked to specific examples or problems shown in the tapes. Phone calls made to other students or the instructor for help may require a student to locate specific sections of the videotape. Unlike live classrooms, which can provide immediate feedback, video and print telecourses may raise questions that take days to resolve.

11.3.2.2 Video and Conferencing Telecourse

This model, characterized by The New School's *Distance Instruction for Adult Learners* (DIAL) program, adds a PC and a modem to the home instructional technology mix. Presentation or taped classroom lecture videos are used in conjunction with printed textbooks and study guides. Unlike the video and print model, instruction is supplemented by computer conferencing–based collaboration with other students and faculty. Questions about course concepts are likely to be answered on line in a more timely fashion than through "telephone tag," but they still require reference to disparate video, electronic text, and print materials.

11.3.2.3 Interactive Video Telecourse

This model, as developed by NYU's Virtual College program, would integrate digital video into a highly collaborative Lotus Notes groupware environment on a single platform: the home PC. All instructional materials—video, lectures, laboratory projects, and readings—are electronic and interactively accessible through one common user interface. In this system, a student having a problem can go beyond the video segment to access directly a hypermedia store of additional information about the topic. If the additional material still does not resolve the problem, the student can capture the appropriate digital video frames, annotate them as necessary, and add them to an electronic message for the instructor and/or other students. Respondents can add and highlight their own digital visual images in trying to clarify the student's problem. The overall process is entirely seamless and highly responsive.

Each interactive video telecourse database organizes information in a nonlinear hypermedia format that supports active cross-references and permits users to jump to various parts of the database as desired. Interacting with the database through associative links, users can follow their individual trains of thought and nonsequentially access video, audio, software, data, graphics, and text materials at varying levels of detail.

11.3.3 Interactive Video in the Home

The nation's telephone and cable companies are committing billions of dollar to build interactive TV systems. Bell Atlantic, for example, plans to provid interactive TV to over 1 million homes by 1995 and more than 8 million b 2000. In designing these interactive systems to lure consumers from malls an movie theaters, however, the telecommunications giants are overlooking a mar ket that is a potential goldmine: the home-based college student.

Unlike TV shopping or movie reruns—often considered by consumers t be free services—college has never been considered free. Average 1992–199 college tuition costs were $2,300 for public institutions and $10,500 for privat ones, and totalled $37 billion in direct tuition payments by students. This $3 billion would put higher education in second place (after catalog shopping among interactive TV's greatest potential sources of annual business revenu The same consumer who might balk at an additional $10 a month for the elec tronic equivalent of going to Macy's might not hesitate to pay a far large monthly fee for gaining access to interactive college telecourses, especiall those offered at lower tuition rates.

As they did with such older media as radio, television, and videocon ferencing, hundreds of colleges and universities will want to offer courses an degrees via interactive TV systems. This new medium will provide homes wit communications and computational capabilities that exceed those avail- abl in most business offices. If the rich instructional promise of this nev medium—far richer than that first offered by broadcast television in th 1950s—is to be realized, work needs to begin on interactive video telecours design and evaluation.

11.3.4 Interactive Video Telecourse Development

To evaluate the instructional potential of interactive video for home-base learners, the Alfred P. Sloan Foundation awarded NYU a $380,660 grant for th development of four interactive video telecourses as part of The Virtual Colleg graduate teleprogram in information technology. The Sloan-funded Interactiv Video Telecourse Project will use the Lotus Notes delivery platform but incor porate many of the capabilities of future interactive TV systems.

The project will utilize two new applications developed for the Lotu Notes groupware package: ScreenCam and Video for Notes. *ScreenCam* is a re cording utility that produces full-screen, audiographic "movies" of PC displa activity. ScreenCam file sizes depend on what is being recorded, but a typica minute of screen activity and accompanying audio occupies only 700 KB o disk space. *Video for Notes* is a video-storage and delivery system that allow users to capture and embed video clips in Notes applications and databases With Video for Notes, students can activate icons located in various telecours lesson, case study, or text documents and have instructional video clips sent from the Virtual College server to their PCs.

The teleprogram will employ 128-Kbps ISDN lines that are capable of delivering digital video into home PCs over ordinary copper phone wires. The teleprogram will be delivered initially over NYNEX ISDN lines in the New York metropolitan LATA. The New York LATA includes New York City and Nassau, Suffolk, Westchester, Rockland, and Putnam counties. The region has a population of 11.2 million with 4.5 million residential phone lines. NYNEX has initiated a "Metro Showcase" program to accelerate ISDN availability within the New York LATA.

Interactive Video Telecourse Project students and faculty will be loaned the necessary ISDN hardware for their PCs during their participation in the two-semester teleprogram. In addition, each faculty PC will be equipped with a desktop videoconferencing system and camera for synchronous curriculum discussion and development. Figure 11.5 shows a basic representation of the project's network and computer delivery system to PCs in students' and faculty's homes.

11.3.5 Interactive Video Telecourse Evaluation

The Interactive Video Telecourse Project will be designed to provide accurate cost and effectiveness outcome measures given a varied range of course session

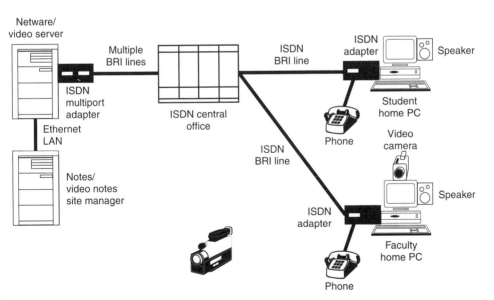

Figure 11.5 Network and hardware for delivery of Interactive Video Telecourse Project.

skill levels—knowledge, comprehension, application, analysis, synthesis, an
evaluation—and telecourse activity/technology processes. The four main tele
course technologies—digital video (full-motion and audiographic), hypertex
computer conferencing, and online laboratory are described below.

- *Digital video* will provide for audiographic and full motion video of fac
  ulty discussions, animated demonstrations, and case study simulation
  The visual linkages created by the video sessions will increase studer
  mastery and retention of telecourse concepts, methodologies, and tools.
- *Hypertext* will provide students with dynamic cross-references to all tele
  course materials and will permit them to jump to various information set
  as desired. Interacting with the telecourse databases through associativ
  links, students will access video, graphics, software, data, and text at vary
  ing levels of detail.
- *Computer conferencing* will support asynchronous discussions amon
  students and faculty in the study of telecourse topics, case studies, pro
  jects, and assignments. The conferences will generate a considerable an
  permanent record of student-faculty interaction and analysis.
- *Online laboratory* will permit collaborative work groups of students an
  faculty to go beyond discussion to creating tangible information system
  products with the tools of the trade. Applications software for computer
  aided software engineering, project management, database managemen
  and quantitative analysis will be supported in the online labs.

The project evaluation will analyze and identify the most appropriate mi
(and corresponding costs) of course activities and interactive video telecours
technologies for meeting the student knowledge and skill objectives shown i
Table 11.4. With over 150 individual topics representing the six student ski
levels in the APC program, the project will provide a sufficient number of case
to reliably evaluate the relative effectiveness (i.e., impact on student skill acqui
sition) of the various technologies. Accurate production and delivery cost dat
will be collected for all telecourse modules to ensure that valid cost-effective
ness analyses of each technology can be conducted.

11.3.6 Education on Demand

The current Virtual College teleprogram uses the seminar model as its basi
instructional paradigm. This approach provides for significant student-facult
(and student-student) collaborations, without increasing faculty costs or reduc
ing scheduling flexibility. However, while teleprogram students can work o
their telecourses at any time of the day and from anywhere, they must still ad

Table 11.4
Matrix of Student Skills, Course Activities, and Telecourse Technologies

Student Skills	Course Activity	Telecourse Technology
Knowledge	Lecture/discussion	Video/hypertext Computer conferencing
Comprehension	Laboratory	Video/hypertext Applications software
Application	Team project	Computer conferencing Applications software
Analysis	Seminar	Computer conferencing
	Library research	Online database/hypertext
Synthesis	Faculty advisement	Computer conferencing E-mail
	Assigned homework	Applications software E-mail
Evaluation	Examination	Online testing E-mail

here to the fixed telecourse session schedule and may take courses only as they are offered on a semester basis (with three start times per year).

The Virtual College now offers instruction the way a school does: on a fixed schedule established by the institution. But there are other educational delivery models. Why not also offer instruction the way a museum does: on a fluid schedule established by the individual? Such a delivery method, coupled with emerging knowledge-agent technologies, might effectively use the independent study model as its basic instructional paradigm. For certain distance learners and for certain courses, that might be an equally effective means of instructional delivery.

The Virtual College will develop an education-on-demand system for delivering quasi-independent study telecourses to home PCs whenever individual students want to begin a course. Each interactive telecourse would have built-in workflow automation capabilities to track, prompt, record, and evaluate student progress through the course. While each student would work independently on the courses, computer conferencing, e-mail, and voice mail access to a faculty and advisor pool would be available to advise, answer questions, and evaluate assignments. Electronic meeting and activity spaces would provide for some degree of student and faculty socialization.

11.4 EXAMPLE OF COURSE ACTIVITIES AND ASSIGNMENTS

The Appendix is the summary of the objectives and assignments for each of the six sessions in the fall 1993 *Systems Analysis* telecourse. During the six-week duration of the course, 21 students, the instructor, and a "client" collaborated on line to analyze, design, and build a working information system in Lotus Notes. As part of this collaboration, over 1,800 Notes documents were created by telecourse participants. For this course, Richard Vigilante was the instructor, Michael Mocciola the client, and Howard Deckelbaum the Notes administrator.

Case Study: Westcott Communications and Distance Learning

12

Westcott Communications, Inc. (WCI) educates, trains, and informs workforces and students in well-defined areas using video-based products for distance learning. The company's strategy is to be the first entrant in a selected market with proprietary network programming and then to rapidly build large subscriber bases. The informational content of the programming is high, developed through the talents and knowledge of industry experts working with the company's creative programming and production personnel. Selected programs are designed to be viewed in sequence as a training course or as a supplement to a training class. A number of subscribers tape portions of broadcasts for viewing at a more convenient time.

Westcott Communications currently has 22 operating networks that provide education, training, news, and information to more than 20,000 subscribers and an estimated 2.5 million professionals and students in the corporate and professional, automotive, financial services, government and public service, education, and health care markets.

The company began in 1986 with one network, the Automotive Satellite Television Network™, which broadcasts news, training, and motivational programs to automobile dealerships across the nation. Current networks include the Fire and Emergency Television Network™, reaching fire and rescue departments; the Health and Sciences Television Network™, reaching hospitals with medical training; the Long Term Care Network™, broadcasting to nursing and

This material was provided by Bryan Polivka, vice-president of programming for Westcott Communications, Inc., and Scott Finley, programming manager for the Westcott Healthcare Teleconference Group, with contributions from Mike Mooney, Dr. Linda Harrington, Jo Streit, and Fred Sylvester. Wescott Communications operates and programs national networks for distance learning that reach subscribers in the workplace and the classroom. This material is provided in order to supply a case study of a business that has been a leader in distance learning since its inception in 1986.

convalescent homes; the Law Enforcement Television Network™, catering to police departments; TI-IN Network®, which delivers K–12 education to schools; the Interactive Distance Training Network™, broadcasting on an ad hoc basis to major corporations delivering personalized training; and the West-cott Healthcare Teleconference Group™ and the Interactive Medical Network Group™, providing ad hoc live teleconferences to health care groups. Other Westcott networks deliver product on videotape, either on a monthly or an as-ordered basis.

Westcott's mission is distance learning, whether the curriculum involves driving emergency vehicles for fire departments or aquatic therapy for rehabili-tation centers or money handling for bank tellers. This chapter looks at this mission from several broad perspectives and then examines several specific examples.

12.1 WESTCOTT'S FACILITIES

The facilities utilized by the company for the majority of broadcast and tape product are designed for maximum quality with extremely high throughput. The primary company studios are located in Carrollton, Texas, a suburb of Dal-las. The majority of WCI productions are staged at an 11-million-dollar facility that includes four production studios, including two 3,000-square-foot studios, one 900-square-foot studio, and one 300-square-foot studio designed for live, classroom-oriented productions utilizing four robotic cameras. All studios are equipped with lights, cameras, and audio equipment. Six additional studio floor cameras are available, all of which can be moved from studio to studio. Included in the studio camera complement is one boom camera mounted on a Merlin crane. All cameras except the boom are equipped with color Q-TV prompter monitors.

Two primary production control rooms are used, each equipped with a Grass Valley 200 production switcher. Each switcher controls over 25 video sources, including all cameras, graphics, still stores, VTRs satellite downlinks, and more. These control rooms also feature individual Dubner Graphics Fac-tory units with paint, each offering a full selection of font styles and sizes, avail-able in over 16 million colors.

The same rooms are used for postproduction editing and are equipped with Grass Valley 151 editors, which command all other control room equip-ment. The high throughput of the facility is tied to the design of these two rooms, the "Big Rooms," in Westcott parlance. All postproduction equipment is routable in real time for use during live or live-to-tape studio sessions. Pro-grams are designed to be finished in the studio, with a host or hosts providing introductions and wraps from a set, with all the high-end graphics and effects carefully used in this venue. Thus, for a great percentage of the programs that are not aired live, no postproduction editing is required.

Additional equipment available to the two Big Rooms includes two-channel DVE as well as a digital disk recorder capable of layering hundreds of video passes to make building effects easier with no generation loss on tape. These control rooms also have access to all major broadcast tape formats, including 1-inch, ½-inch Beta and Beta SP, ¾-inch, SVHS, and digital D2. Fifteen Sony BVW-75 Beta SP VTRs are also available, featuring slow motion and reverse effects. The technical support area also features a Grass Valley Horizon routing switcher, allowing instantaneous connection of any audio-video source to any destination for editing, dubbing, or recording. They are equipped with a Yamaha 3000 audio mixer, which can handle up to 24 audio sources at once. Each room also features playback facilities for cassette tape, NAB cartridges, reel-to-reel tape, phonograph records, and CDs, in addition to videotape formats. A voice booth with Neuman microphones is available to either control room.

Control room C is separated from Studio C by a glass wall and integrated into its overall design for classroom-type presentations. The studio houses four robotic-controlled cameras operated from the control room. These cameras are equipped with special shutters that allow them to shoot computer monitors without annoying flicker. Control C also features a Grass Valley 100 switcher, a Grass Valley 131 editor, Grass Valley AMX 170 audio board, Yamaha 12 input board for studio production, Abekas A53D single channel DVE, and Dubner 20K character generator, and each camera is equipped with Q-TV computer-generated prompter.

Four Sundance logging stations are available for producers, as well as five cuts-only editing bays. Westcott Communications has also invested in non-linear editing in the form of the Avid Media Composer and Airplay products for use by producers. The building also houses one integrated production system (IPS) suite, equipped with the Grass Valley Integrated Production System. That system features a Grass Valley 100 switcher, Grass Valley VPE-141 editor, AMX-170 audio console, and a Dubner character generator interfaced to three Sony BVW-75 decks. The logging stations, cuts-only editing, and IPS systems are all used to create "packages" that can be included in long-form programs live or live-to-tape.

The Carrollton facility has two satellite uplink systems, including two 7-meter Scientific Atlanta KU band uplink systems with redundant transmitters, capable of clear and B-Mac encrypted analog transmissions as well as compressed digital video transmissions.

To ensure that all networks run smoothly, WCI employs a full-time, round-the-clock staff of television, videotape, and satellite communications experts. Customers on WCI satellite transmissions can call a toll-free number whenever there is a problem in program reception. The company also provides for each of its satellite networks a magazine-style program guide featuring the daily lineup of programming.

Bringing in the vast quantities of video used in production of programs are six field production crews constantly crisscrossing the continent. WCI also makes extensive use of an inhouse video library that currently includes over 76,000 titles, the vast majority of them produced at WCI.

12.2 DELIVERY SYSTEMS

Westcott Communications is committed to staying on the leading edge of technology, but it is a user rather than a creator of technology. All decisions about investment in advanced technology, whether in production, delivery, or storage of product are made on the basis of careful analysis of costs versus benefits to the customer and to the company. In most cases, the technology selected will be that which is most common and most accessible to customers and potential customers. The primary exception to this policy in the past five years has been in the delivery area. In 1993, WCI made a strategic decision to deliver its satellite product via *compressed digital video* (CDV) and became the first entity to utilize that technology on a large scale to deliver commercial television product.

The primary factors that weighed in the decision were financial—the company at the time was leasing one transponder for each of its broadcast networks. The cost reduction capable of being realized by putting five networks on one transponder justified both the expense of migrating thousands of receive sites to CDV and the risks of being the pioneer with technology that had been beta tested but not field tested to any great extent.

After thorough investigations and negotiations, WCI elected to utilize product supplied by Compression Labs Incorporated. Currently, all Westcott networks (with the exception of ASTN®) broadcast using the CLI Spectrum-Saver technology. LETN®, FETN®, HSTN®, LTCN®, IDTN®, and the health care teleconference products of WHTG® and the IMN Group™ are 6.6 Mbps, while the three channels of TI-IN Network product are 3.3 Mbps. ASTN, the automotive network, continues to use standard analog broadcast because it is the only network in which any piece of the ground segment, the satellite dishes on the roofs of auto dealerships, are not wholly owned by Westcott Communications. All General Motors and Chrysler dealerships are equipped with VSAT dishes providing two-way data links between the dealerships and the manufacturer. Westcott Communications enjoys long-term lease agreements with both manufacturers, allowing the network to switch on those dealerships as new subscribers from the Carrollton facility without any installation costs. These broadcasts are BMAC encrypted.

12.3 WESTCOTT PRODUCTION TECHNIQUES

Westcott Communications produces over 130 hours of continuing education and training product on video each month from its Carrollton studios, most of it created in a format designed to mirror the production values of standard network television, rather than a classroom. (An additional 540 hours of live classes are broadcast each school month on three channels originating from the San Antonio Teacher Work Stations at TI-IN Network). WCI's overall strategy for creating training–continuing education product has been to hire top-quality video producers who have shown some predilection, however informal, for education and teach them basic instructional design, such things as establishing measurable objectives and teaching to those objectives. Most WCI products include testing, syllabi, or other collateral materials, which are also the responsibility of the producer, under the supervision of a director of education for that network. It is generally the responsibility of the content expert, however, to provide the instruction. Therefore, the producers at WCI must be adept at working with content experts, faculty, and trainers to transfer successful lesson plans and instruction from a classroom or seminar environment into a satisfying video product.

In 1986, WCI created all products for the Automotive Satellite Television Network in a talk-show format, with a host on a living room–style set introducing a speaker, chatting with him or her for a few minutes, then letting the content expert stand up and deliver the content. These programs were created in a high-volume, low-cost style, so most of the dollars could be spent on getting the biggest-name motivational speakers to the studios, for example, Zig Ziglar and Anthony Robbins. This was Westcott Communications' "era of mass production." As many half-hours were shot in a day as could be squeezed into the schedule, with 10 to 15 programs per day not unusual. As is the case with any mass production operation, there were quality issues—some resolvable in editing, others not. For every 10 programs taped, one to two were scrapped as unairworthy.

In 1988, the company introduced a new style of production to its ASTN product: scripted, postedited, actor-portrayal products. This began a period of time that might be called WCI's "craftsmanship era." A series of 26 programs entitled "Customer Satisfaction Management" became the cornerstone of the new Westcott look. When the Law Enforcement Television Network was added in 1989, careful preproduction, film-style reenactments and high-end editing and graphics carried the day. Each individual half-hour program stood on its own, and the production style mirrored prime-time drama. Hollywood producers were hired, and a former Miss America and a national commercial actor hosted the flagship LETN program. But WCI quickly realized that, while the glamour was appreciated by many, customers generally signed on for hard-core, meat-and-potatoes training; if they didn't receive it, they weren't satisfied regardless of the production values.

In 1991, a new approach was taken with the issuance of the Westcott Communications *Stylebook*, an internal manual that focused on carefully defining the learning objectives and building at least one achievable objective into every 7 to 10 minutes of product. Coupled with that was a conscious effort to produce series of programs, five half-hours on a single topic, which allowed the same preproduction hours to stretch much further. The goal was to produce a month's worth of segments in the field during one shoot lasting only a day or two and then to finish those four or five programs in one studio session. WCI again disdained heavy postproduction but still brought to bear highly polished video elements in opens, transitions, teases, and graphics. The new process was dubbed "lean production," after the buzzword in the automotive industry at the time, and had the effect of improving quality—as judged by customers—while at the same time dramatically improving business efficiencies. Programs produced in this style resemble high-value, high-volume programs such as Entertainment Tonight or nightly newscasts. The field segments, which carry the meat of the training, are logged and edited offline, and a studio script is prepared to tie them together. In a studio, a TV-savvy host who can read a Q-TV prompter and follow time cues introduces each segment, sets up the learning objectives, and wraps up the segments. The implementation of this process allowed WCI to drive huge amounts of product through the facility without heavy investment in new capacity.

In 1994, the addition of a wide array of health care teleconferences and the *Interactive Distance Training Network* (IDTN) launched the company into a new era of live production, forcing yet another change in the production strategy of the company. Live-to-tape sessions took a back seat in the scheduling of studios to the new, high-quality live programming. Alternative approaches were explored—off-hour studio sessions, desktop editing, field-intensive, post-edited pieces. The live product enjoyed a huge success, particularly with the corporate clients of IDTN, because those customers benefitted from Westcott's years of experience in creating satisfying video product designed to achieve specific, measurable objectives. The Fortune 500 clients were paying for the results—test results, favorable opinions, and responses—rather than just for a production.

12.4 DAILY SATELLITE BROADCAST NETWORKS

The primary business of Westcott Communications during its first decade has been the subscription networks that broadcast from 8 to 24 hours a day, five days a week, into workplaces across North America. The Automotive Satellite

Television Network was the first Westcott network, but it is not the oldest. Pre-dating ASTN by two years was the *Health & Sciences Network* (HSN), which broadcast to hospitals. It was acquired by Westcott Communications in 1992, and its subscribers were transferred into the new *Health & Science Television Network* (HSTN). All the daily broadcast networks—ASTN, LETN, FETN, HSTN, and LTCN—share a similar structure and mission. Executives and senior officials at local sites in each field—auto dealerships, law enforcement agencies, fire and emergency service providers, hospitals, and nursing homes—make the decision to employ the network to supplement or replace current education and training efforts, agreeing to a monthly cost that may range from $200 to more than $3,000 per month, depending on the market and the size of the institution. The subscribing site designates a site coordinator, who is responsible for utilization of the products and services at that location. That person receives monthly mailings, including program guides, collateral materials such as tests and syllabi, and other informational pieces. The program guide contains schedules and synopses for the programs that will be broadcast during the month. It is the aim of the network to provide in this manner high-quality, low-cost training and education that can be used by a wide variety of people at that location.

The Fire & Emergency Television Network, for example, provides training curricula for emergency medicine, firefighters (levels I and II), hazardous materials response, chief officers, first-line supervisors, rescue, and industrial. Each of these is based on the requirements of the *National Fire Protection Association* (NFPA), and each program references a specific NFPA standard. Collateral materials include directions and information on how to provide hands-on training for skills related to and demonstrated in the video, so that the training officer at that location can gather up the needed equipment and conduct the subsequent skills-based training sessions.

Health care providers in hospitals and nursing homes who receive regular continuing education programs include nurses, doctors, technicians, operations employees, food service workers, and supervisors. In law enforcement, regular training reaches patrol officers, investigators, sheriffs, corrections officers, supervisors, and dispatchers, with programs touching other specialties from time to time. In the automotive area, salespeople receive the lion's share of programs, while other audiences include sales managers, parts department managers, service advisors, service technicians, finance and insurance managers, receptionists, and the business office.

Industry-specific news has always been a large part of these networks, generally provided in five-minute daily newscasts interspersed between training and education programs. These products are created inhouse, from video supplied under contract with major news organizations and acquired by inhouse reporters and photographers. While keeping the viewers apprised of developments relevant to their work and their careers, this programming offers a network feel to the product, generally produced in a manner consistent with the network evening newscasts, with which viewers are familiar.

12.5 VIDEOTAPE NETWORKS

The *Professional Security Television Network*™ (PSTN) was Westcott's first videotape network, launched in 1990 for professional security officers and managers. It was initially designed to be broadcast, but difficulties with building permits and costs associated with providing downlink equipment and running cable through high-rise buildings, where many security operations reside, forced a new strategy: delivery of the same product on tape as a monthly subscription. Almost concurrent with the launch of PSTN® was the acquisition of American Heat Video Productions, Inc., a St. Louis–based company that provided firefighters a monthly video news magazine replete with fire and rescue and inside stories with lessons learned. Shortly afterward, *Pulse*, for emergency medical providers, was launched. Since then, videotape subscription services have been launched or acquired which service workers in accounting (CPA Report™ and the Accounting and Financial Television Network™) and in local government (Governmental Services Television Network™, a partnership with ICMA, PTI, National League of Cities, and National Association of Counties).

Also included in the Westcott Communications videotape networks are business units that sell individual products, either through catalog sales or direct marketing. These include two entries in the field of industrial safety and training, both acquisitions (Industrial Training Systems Corporation and Tel-A-Train, Inc.), and one in general business/management (Excellence in Training Corporation).

Bankers Training & Consulting Company, also an acquisition, is a "videotape" network that sells videotapes, computer-based training, CD-ROM, and other products to the banking industry. These products typically are not sold as individual items but rented as part of a contractual agreement that allows customers access to the whole library of products. Customers order the programs, use them or view them, then send them back and order more. This training-on-demand, library approach has been adopted and used as an add-on service in several of the satellite-broadcast networks, including LETN and FETN.

12.6 INTERACTIVE DISTANCE TRAINING NETWORK

IDTN utilizes Westcott technology to provide turn-key services, including course conversion, instructional design, studio and field production, and training through 44 strategically located electronic classrooms across the United States. IDTN minimizes the time and costs associated with moving large numbers of people to a common training site. Major U.S. firms using IDTN include Oracle Corporation, Intel Corporation, Compaq Computer Corporation, Eli Lilly Corporation, Sun Microsystems Computer Corporation, and Storage Tek, as well as Glaxo Pharmaceuticals and Whirlpool.

By using IDTN and its distance learning concept, the specific training, education, and conferencing needs of those companies are met. Computer product reseller training for field representatives enabled one company to launch sales weeks earlier than expected. Another company cited the need to quickly and consistently educate their representatives on PC usage as the reason for using IDTN. Other applications include the delivery of product training, new product announcements, hands-on computer training, sales training, and corporate communications with thousands of employees, representatives, and administrators.

IDTN broadcasts originate at Westcott's Carrollton studios and are beamed live to the electronic learning and information centers. The centers are leased by and controlled by WCI and have a 24- or 36-seat capacity, with a total nationwide capacity of 1,180. The centers are convenient to airports and hotels and are equipped with two 27-inch monitors and a microphone-equipped response keypad at each desk.

The broadcasts are one-way video, two-way audio to the sites, with three-way audio capability (from students in one classroom to students in another). Programs are secured by an encryption process. Responders use the keypad to ask questions or talk back to the broadcast presenter. IDTN presenters can address participants by name, engage them in discussions on subject matter, and get real-time feedback. Participants can ask questions, alter the pace of instruction, answer quizzes, and respond to other participants at other IDTN sites.

Data transmitted through the keypads at the electronic learning and information centers allow instructors in the studio to immediately measure the effectiveness of their instruction. Additionally, data are collected and stored for further evaluation after the session.

The keypad system, developed by One Touch, can issue the following types of questions in real time:

- True/False;
- Yes/No;
- Yes/No/Undecided;
- Multiple choice (A, B, C, D, or E);
- Numeric (requires a numeric entry on the keypad).

Formatted questions can be used throughout the presentation to test or review information or provide a quick baseline of current knowledge among the participants. Questions can be presented alphanumerically, with a picture/graph, or both. Questions may be displayed followed by the results as they come in. A final display can then highlight the correct answer. On issuance of the questions, the host system displays a graphic response screen. A bar chart builds, indicating all responses to the questions, once again in real time. The information can then be displayed to the entire group. The response screen indicates the total number of responses to the question, how many individuals responded to each answer option, and the percentage of responses to each answer. Once a student has selected a choice for any of the questions, other than a numeric question, he or she will not be able to change that response. Numeric entries can be cleared and reentered prior to sending. For numeric answers, the response screen used by the presenter will indicate the total number of responses, the highest number entered, the lowest number entered, the total sum of all numbers entered, and a mean score. A presenter also has the option to issue pop questions, which can be issued in all of the aforementioned formats.

In addition to the interactive questions issued in real time during an event, the system also has the capacity to issue comprehensive tests and quizzes. These tests can be developed and delivered to students on paper before, during, and after an event. They can take the form of preexaminations or postexaminations that analyze knowledge or performance gains. Tests are different from the other interactive questions in that they require the issuance of hard-copy questions and are not presented on the site location monitors. The exams are still taken on the interactive keypads, and the host system will still collect and analyze all data. As with numeric responses, students completing a test have the option of checking their answers and making changes. When they are satisfied with all entries, or if time has run out, students respond to the last question of the test, which asks whether they are finished. The number of students finishing their exams will be displayed on the response screen in the host system, letting the instructor know when everyone has finished.

Other features of the IDTN site center keypads include a call button, allowing students to "raise their hand" to ask a question or make a point. When activated, a red icon depicting a student with a raised hand, along with the student's name and location appears on the host screen. If the instructor wants to speak to the student, touching the icon with a finger activates the student's microphone and allows a direct conversation. An additional student can be added for a three-way call. Preselected subject matter experts, who may be present at several different downlink locations, may appear as "expert" icons (a student with a mortar board). This identification serves two purposes. First, the moderator or presenter will quickly be able to identify if a colleague is attempting contact; second, the host system treats expert icons with a priority, jumping their responses to the head of the call line when several students are waiting to talk. This allows expert colleagues to respond in a more timely fashion to discussion.

The host system also has the capability to issue a random call to students. When this icon is pushed, the host randomly picks a keypad and activates a student's microphone. It is a quick spot-check to verify that students are paying attention. The presenter can also call any student at will. It generally takes about 30 to 45 seconds to ask a question, have the students respond, and display the results.

IDTN essentially allows every person at every site to receive the same message at the same time and puts a company's best trainers in front of the entire company organization at once.

The basic IDTN package includes the use of the learning and information centers, use of the interactive One Touch system, satellite transponder space, production facilities and staff, rehearsal prior to the program, three to five hours of broadcast time, instructional design services, operational support, a master videotape, site facilitators, and data collection and reporting. In their instructional mode, presenters can move freely around the studio, and the interactive host computer system can be operated by IDTN personnel or the presenter.

Optional services by IDTN include creation of opens, roll-ins and graphics, custom-built sets, the capability of installing temporary downlink sites with origination from the remote site possible, live video from the remote location, the availability of international transmission, and specialized reporting.

12.7 TI-IN NETWORK

Another major area of distance learning at Westcott is the TI-IN Network. TI-IN offers programming in seven different areas: student instruction, staff development, student enrichment, student test reviews, special programs, subscriber training, and the CNN classroom. During the 1994–1995 school year, TI-IN offered high school credit for courses in math, science, foreign languages, social sciences, and English language arts. Elementary and middle school courses in foreign language, math, and science are also offered. The network also offers courses designed to prepare students for the Scholastic Aptitude Tests (SATs) and ACT.

The TI-IN Network uses the same One-Touch system in place at the IDTN sites. Each TI-IN site is asked to assign a facilitator for high school credit courses, a teacher partner for elementary–middle school courses, and a TI-IN academy administrator for staff development sessions. These individuals are responsible for monitoring and assisting in the administration of TI-IN network programs, much as site facilitators are responsible for overseeing the taping and distribution of programming from other WCI networks. Each TI-IN site may assign as many local personnel as deemed necessary. For example, one facilitator may be assigned to a Spanish course, another to a French course, and another may be the TI-IN academy administrator for staff development. TI-IN requires each high school site to assign a facilitator who remains in the classroom during the entire instructional period for each credit course in which a site registers students. Facilitators are responsible for many duties, including monitoring class attendance, collecting schoolwork, distributing materials, and monitoring tests. The facilitator should also keep the school principal updated and informed of student progress.

TI-IN requires each site to assign a teacher partner for elementary and middle school courses. It is not necessary for teacher partners of foreign language classes to know the language or content being taught. However, the teacher partners should be willing to learn along with the students. Before class begins, teacher partners should prepare students to watch the class by focusing the students' attention and making sure that all materials are available. As with facilitators, teacher partners remain in the class during the instructional period.

To measure viewer response, TI-IN uses the interactive keypads, one per student. For students who require supplemental instruction, tutoring is available through toll-free telephone lines and on-air tutoring sessions. Generally, students may speak with the teacher or teaching assistant before, during, and after class. Each TI-IN instructor has a daily conference period and weekly evening office hours. During those times, students may call a toll-free student assistance number and speak to the TI-IN teacher or teaching assistant. Each site is responsible for securing the materials and equipment identified for the student instruction courses. The required textbooks, lab apparatus, resource materials, workbooks, language audiotapes, and other materials for each course are listed in a course description provided to the sites. The site facilitator is responsible for distributing materials to each student as instructed by the TI-IN network teacher.

The TI-IN producer at Westcott works primarily with three categories of on-air presenters: classroom teachers, professional development consultants, and student enrichment presenters. The extent of production support required varies from individual to individual and is determined to a great extent by the subject matter that is to be presented and the approaches taken by the presenter. Production techniques and methods usually are not well understood by the presenter and require that the producer be an excellent communicator who can gather critical insights for the presenter and then creatively translate those ideas and concepts into visual and/or auditory images that will reinforce the ideas being taught.

TI-IN teachers prepare lesson plans, develop teaching strategies, and produce computer-generated graphics to support their classes. Producers work with the teachers in the capacities described above. They are required to understand and promote the objectives of the instruction.

Producers for professional development are responsible for selecting timely topics that are relevant to educators and also for selecting, contracting with, and scheduling professional development consultants. Professional development consultants typically require a different level of support than classroom teachers do. Their presentations usually are very structured and organized for in-person seminars or conferences. That requires that producers thoroughly research the program topic first and then take the presenter's prepared materials and organize and translate them into a logical sequence and format for television. Additionally, producers need to synthesize the presenters' program objectives and to develop visual support materials appropriate to their presentations.

Student enrichment presenters usually have very good experiences and ideas but little in the way of organized formats or visual support for their programs. Producers need to develop ideas for field shoots and graphics preparation to help these programs be successful.

12.8 M.B.A. PROGRAM

Westcott Communications also offers the first college degree program for subscribers to its HSTN. The M.B.A. program has a focus on health services management. Offered by the University of Dallas, the M.B.A. program in health services management is designed to give added breadth and depth to professionals serving in administrative capacities in hospitals, long-term care facilities, clinics, and other health-related organizations.

Courses delivered by an instructor from the Carrollton studios include Financial Accounting, Strategic Marketing in Health Care, Managerial Accounting, Statistics, Monetary and Fiscal Policy, Hospital Administration, Health Insurance and Managed Care, and Financial Management for Health Services.

The program is designed for mature students with management experience. It does not emphasize background or theoretical courses but a highly pragmatic program focusing heavily on the realities of managerial life.

The program provides students with the opportunity to tape broadcasts for later use and interact via collaterals and test questions. In that sense, it follows the basic concepts of distance learning but uses Westcott technologies to make it easier.

The talent/instructor faces two cameras, one with a medium shot showing part of the desk and the other with a tighter bust shot. Both cameras are equipped with Q-TV prompter. An additional camera mounted overhead looks down on what the instructor is writing. The talent has control of the prompter scroll rate and direction.

The director/producer has control over the cameras by means of remote controlled pan/tilt heads with remote focus and zoom. The director also has access to the CCU functions of the cameras for iris, black balance, white balance, and setup. The director also operates the character generator, video switcher, audio console, videotape, and edit functions of the prompter.

The use of the servo-operated robotics described earlier with multiple-shot memories, as well as graphics computers and character generators with sequencing capabilities and switchers and audio consoles with programmable settings, helps the director take full advantage of preplanning in setting up the instructional program. The talent/instructor has only to think about the material to be presented and the proper, energetic presentation of that material. To date, the M.B.A. series has been very successful, setting a new standard of managerial competence in the increasingly complex health care arena.

12.9 THE WORKSTATION AND INTERACTIVE MEDIA

Westcott Communications' view of the future is interactive. However, for WCI the advent of interactive recorded media such as CD-based products is a two-edged sword. Many of the company's happiest customers in fields such as law enforcement, fire and emergency, and, to a lesser extent, healthcare have libraries full of WCI products that they would prefer to continue using, and businesses in general are more hesitant than individual consumers to pay for the latest technology without a good business case. Long, expensive product-development cycles for untried media are not the foundations on which Westcott Communications was built. But clearly, the end of the VHS tape is in sight. So rather than invest heavily on a questionable return, WCI spent 1993–1994 developing an interactive workstation that featured a high-powered PC, a video card, and proprietary software that allows network customers to organize all their tape or broadcast-based training and lets individual users interact on a sturdy, proven, user-friendly computer equipped with a mouse and a modem. The LETN workstation, called the STTAR system, comes loaded with every test and every piece of collateral material that accompanied every program created for the network in the last several years. No matter which tape is pulled from the library, no matter how any idiosyncratic subscriber may file tapes and handle training records, all the user needs to do is to sit down at the computer, enter his or her ID number, and enter the program number; the test for that program then appears on screen. The tape is played through the computer screen by use of the video card, and the viewer then uses the mouse to take a 10-question, multiple-choice test. The test is graded on the spot, the viewer is remediated where appropriate, and the results stored in the computer. All such records can be accessed by a training officer with a password, printed locally, and uploaded via modem to Westcott Communications, where the Interactive Education Services department keeps the required third-party documentation, and can send a certificate of completion to viewers who deserve one. The same modem downloads new course information and tests every month to the unit.

The workstation manages not just LETN training but any and all training done at that department. A Test Creator program allows a training officer to create and store tests for new products or rework LETN's tests to fit local law or departmental policy. It is a one-stop training management system that bridges the past—volumes of tapes—with the future. Whatever products WCI creates in interactive formats, on whatever platforms, they can be integrated into the STTAR system by a simple hardware upgrade.

12.10 DISTANCE LEARNING GUIDELINES

Distance learning at Westcott Communications also involves video teleconferences. Participants can call in on toll-free lines to communicate directly with presenters and faculty during the real-time run of the program. Videoconferences range from presentations with industry leaders to a chance to go one on one with some of the nation's top motivational speakers. Continuing education credit is offered for almost all WCI products.

As discussed in the section about TI-IN, the key to Westcott products is the producer. The producer bears the responsibility for scheduling many things, including in-house graphics time, edit time, field shoots, scripting, crewing the shoot (studio or field), and working closely with the content expert. To that end, WCI provides producers with several stylebooks and programming creation guidelines to help them reach their goals.

Westcott Communications uses an approach to distance learning that combines education and entertainment to yield a satisfying educational product. The use of this method of instruction is appropriate and important in capturing and maintaining the students' attention from a distance. The benefits of this approach are broken out in the two basic tenets of WCI programming: education and training.

Though various markets define the two terms differently, sometimes using them interchangeably, it has been useful for WCI to define education as a more formal approach to instruction than training, with more formal results in terms of diplomas, degrees, and continuing education units. In both education and training, products are based on learning objectives, which in turn are based on program objectives, usually derived from a curriculum. Education is more cognitive or thought oriented in nature, meaning the thrust of the instruction is to increase knowledge and comprehension, while training tends to involve more behavioral objectives.

While producers provide the "magic" that ultimately develops into a satisfying visual and educational product, the faculty generally provides the meat and potatoes of the product. In health care, for example, a content provider will supply learning objectives, program synopsis, detailed outline of the topic, list of current references, and 10 posttest questions derived from the learning objectives. These will be compiled into a syllabus with a test and, added to a 30-minute video program, will provide 1 hour of educational activity and thus one continuing education unit.

Each program from WCI generally begins with a statement of the learning objectives. This is done with voiceover from the faculty or someone else and graphics. Learning objectives differ from program objectives. Program objectives provide an overall goal of the program, whereas learning objectives describe what the students should be able to learn from the program, that is, what they will be able to do after viewing. WCI producers are encouraged to give viewers something to do, not just something to learn.

Video is, of course, an important part of any WCI production, but as in broadcast news, the producer is encouraged to refrain from "wallpaper" video—video for the sake of video. Instead, any video used should be able to provide demonstration. Training segments on WCI programs, in the 30-minute programs, tend not to be longer than 10 minutes, with an ideal length of four to seven minutes. This length is a nod to the realities of commercial television and home viewing habits: Viewers start to drift and to lose focus without a break every so often. By focusing on and accomplishing a smaller, enabling objective in this time frame, the demands of both television and education are satisfied.

Each program also concludes with a summation of the learning objectives. This summation serves two purposes: It reminds the student of the material covered and reinforces that the content has been covered.

While these may appear to be basic building blocks of any distance learning format, they are also applicable to larger-scale formats such as IDTN and TI-IN. For instance, the program objective is the starting point for any program. It can be a simple sentence, a descriptive paragraph, or a short list of specific information that outlines what the presenter wants to achieve from the broadcast and the outcome expected in the learner.

Once the objective has been defined, program content can be assembled. The producer starts by asking what the participants need to learn to achieve the program objective. The best option is to involve the subject matter expert in this phase of content development. Time must also be considered by the producer, including preproduction time, rehearsal time, production time, actual event time, and postproduction time. WCI stresses heavy preproduction time, since programming facilities are in heavy demand among the networks and programs should be able to enter the production facility highway and immediately pass through to their destination, the viewer. Costs are always a prime consideration as well, because each program has a budget for variable costs such as tape, travel, and talent and a separate budget for production facility charges, such as studio, edit, and graphics. These costs and budgets are managed at the team level, with the programming manager of each network responsible for day-to-day decisions concerning the costs and accountable for the cumulative results versus budget month to month.

The unique qualities of IDL require that certain elements be utilized and incorporated into the learning event. These elements include, but are not limited to, the interactive student response system (used with IDTN and TI-IN), graphics, video, and sound. The primary consideration, of course, in determining the usage of interaction, is: What type of question/answer will best achieve the objectives? This same question must be asked during every phase of production, every decision about special effects, B-roll, demonstration versus explanation, graphics—everything. Three additional considerations that should be made during development of an interactive distance training event at WCI are:

- What type of questions will elicit and sustain learner interest?
- With what frequency should interactive response questions occur?
- Where in the program should response questions occur to sustain interest and maximize learning?

Producers should use interactive questions in the following locations to enhance the appeal of interactive distance training:

- To gain student/participant attention, use frequent student response system questions to present novelty, surprise or unexpected events in the instruction.
- To focus attention, provide prequestions that act as advance organizers.
- To enhance interaction and facilitation and to increase retention, allow for some student response system activities that provide participants with opportunities to construct responses to questions.
- Design programs and activities that maximize interpersonal interaction among students.
- To facilitate two-way and three-way interaction, encourage students to question the presenter, who should respond with clarification and elaboration.

12.11 LOOKING AHEAD

WCI's goal is training business and other professionals; the medium (for now) is satellite TV and tape delivery of products. WCI management is exploring new avenues and new technologies, including the PC-based Workstation, online services, CD-based products of many descriptions, and network delivery of training and educational products into virtually any work environment through IDTN-style classrooms.

No matter what format is decided on, the role of the producer in distance learning will remain unchanged: the creation of product with measurable results. It is WCI's belief that distance learning will continue to grow as a market not only in the United States but on an international level as well. The advantages of distance learning over physically transporting staff and material thousands of miles is seen not only in the ledger column but also in the speed with which personnel can be trained and in the consistency of that training. In addition, the innovations developed for distance learning, such as the interactive keypad, enter into the teacher-student interface directly. A skilled teacher can utilize these tools to create a closer link with a student than would be possible in a live classroom, with less emotional baggage such as peer pressures, fear of failure, or fear of appearing ignorant. In other words, distance learning is rapidly becoming an improvement to the educational process, regardless of distance. But even at its current pace with current abilities and technologies, requirements for continuing education among professionals are increasing and will continue to increase. This will drive the need for more and better training at less cost and in less time.

Appendix:
Example of Course Activities and Assignments

The Virtual College
Course Session 1

Y52.1100 Systems Analysis
Planning the Consortium Systems Project

Analysis   Design   Development

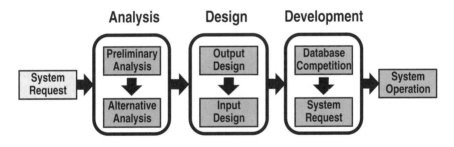

SESSION OBJECTIVES

This teleprogram (see Chapter 10) is a six-week online collaboration during which students and faculty will identify needs, ask questions, conduct analyses, and build a working information system using the Lotus Notes groupware package. Students are expected to read the assigned books and articles, and to actively participate in the online project discussions and analyses. As a systems analysis background is not a prerequisite for the course, the technical depth of student contributions to the project discussions need only reflect the knowledge gleaned from the readings (and the discussions themselves). Throughout the course, students will be asked to contribute suggestions, alternatives and recommendations, and these will be the basis for sixty (60) percent of the final course grade.

A main objective of this first **Course Session** is to become familiar enough with the Lotus Notes groupware package to use it as an effective collaboration tool for our systems project workgroup. This teleprogram will be an experiment for all of us, and there will be times when not everything may work properly. But don't despair! If you've gotten this far in the system, you're more than halfway home. If you run into any problems, please tell Howard Deckelbaum or me. The only way we will be able to correct any problems is if we are told when they happen.

SESSION ASSIGNMENT (DUE OCTOBER 15)

Readings: Kendall and Kendall, Chapters 1 ⬜ and 2 ⬜ (electronic)
Sproull and Kiesler, Chapter 1
Applegate, et al. article ⬜ (electronic)
Overview of The Computer Consortium ⬜ (electronic)

Most of this first week will be spent reading through the various texts and articles you received, and generally becoming familiar with the Lotus Notes package. As an *initial* assignment due on **October 13** (there will be another assignment related to The Computer Consortium systems project that will be due on **October 15**) to test both the teleprogram system and your ability to communicate with it, I would like each of you to compose and send a response to this **Course Session** document. Make the subject of your response (approx. 100 words) a brief introduction of yourself to the other members of the workgroup and a short statement of what you hope to get out of the teleprogram during the next six weeks. Procedures for conducting a database exchange are described below.

Conducting a Database Exchange

A Database Exchange is the process used by Lotus Notes to update all user copies, or replicas, of the databases shared in common with the server. During a database exchange, you instruct your PC to dial the NYU server to exchange new, edited and deleted documents in the databases. All new and modified documents created by you since the last time you logged-on are sent to the server, and all new and modified documents created by all other course participants are sent to your PC. Once all document exchanges have been made, your PC is automatically logged-off the NYU server. At any point in time, all telecourse participants will have current (or almost current) copies of all course databases on their PCs. The more frequently you conduct database exchanges with the NYU server, the more current your personal copies of the telecourse databases will be.

1. Press the **ESC** key twice to return to the workpage. Notice that **The Virtual College** icon is the only database that is highlighted (and therefore is the only one selected for a database exchange).
2. Click on the **File** menu, then select **Database** and **Exchange**.
3. In the **File Exchange** dialog box, select the Server **NYU**, and then select the options **Selected databases, Receive documents from server**, and **Send documents to server**.
4. Be sure to select **Hangup when done**. This will free-up your phone line and a server port when your database exchange is completed.
5. Select **OK**. To the question **Make phone call to server NYU on port COMx?**, check to insure that the COM port is correct and select **Yes**.
6. Notes will then dial the server at NYU and conduct the database exchange. Your response will be sent to the server, and any other responses or updates will be sent to your PC.
7. After the exchange is completed, the **File Exchange Statistics** dialog box opens to give you details about the databases processed. Select **OK** to close the dialog box and end the exchange.

I look forward to reading your responses on or **before October 13.**

The Virtual College
Course Session 2

Y52.1100 Systems Analysis
Preliminary Analysis of the Silicon Valley System

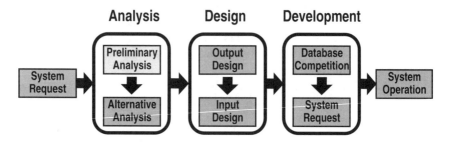

SESSION OBJECTIVES

The main objective of this session is to prepare a preliminary analysis of the current Silicon Valley information system. Within the typical development life cycle of an applications system, the preliminary analysis is usually the first step. The preliminary analysis begins when someone within the organization perceives a problem, wants an existing system modified, or seeks an entirely new system. Preliminary analyses may be initiated by a formal or informal request of senior management, the systems department, or as in our case the system's users (via the Analysis Request Form).

While preliminary analyses vary in scope and detail depending on organizational requirements, the format that our workgroup will use is fairly generic. Our preliminary analysis will include the following sections:

1. **Introduction**
   Basic background of the organizational setting and the application or need being investigated. It may include organization charts or other representations to clarify overall structure and participating units.
2. **Objectives of the Study**
   Brief statement of the current problems or needs.
3. **Problems with the Current System**
   Detailed itemization and explanation of the significant shortcomings with the present system and its procedures, and the respective impacts

on operational effectiveness or efficiency, information accuracy, timeliness, etc.

4. **Costs of the Current System**
   Identify both the annual operating and projected five-year costs of the current system.
5. **Recommendations**
   A recommendation on whether to proceed with the analysis to the next step in the life cycle—preparation of the Alternatives Analysis. If continuation of the study is recommended, expected (and quantitative) goals of the new system should be given.

SESSION ASSIGNMENT (DUE OCTOBER 25)

Readings: Drucker article ▯ (electronic)
Sproull and Kiesler, Chapters 2 and 3
Kendall and Kendall, Chapters 3 ▯ and 6 ▯ (electronic)
Case Study: Preliminary Analysis of the Database System ▯ (electronic)

Our main task for the next ten days is to collect the fairly-detailed workload, procedural and cost data necessary to prepare the preliminary analysis of the current database system. In the interests of time, the workgroup will be divided into four subgroups—Group A, Group B, Group C, and Group D—with each subgroup being responsible for one aspect of the preliminary analysis' data collection. Each subgroup will conduct its work within its own Notes database. Workgroup members have been assigned to the subgroups as shown in Table 1 below.

Table 1. Composition of Project Subgroups

GROUP A (groupa.nsf)	GROUP B (groupb.nsf)	GROUP C (groupc.nsf)	GROUP D (groupd.nsf)
Student 1	**Student 7**	**Student 12**	**Student 17**
Student 2	Student 8	Student 13	Student 18
Student 3	Student 9	Student 14	Student 19
Student 4	Student 10	Student 15	Student 20
Student 5	Student 11	Student 16	Student 21
Student 6			

Each subgroup has an associated Notes database (groupa.nsf, groupb.nsf, groupc.nsf, and groupd.nsf) and members will have access only to the group database that they are listed under as shown in Table 1. Mike Mocciola,

Howard Deckelbaum and I have access to all four databases. Over the weekend, all members will copy their groupx.nsf database from the NYU server to their PCs following the instructions provided below. I am asking each member whose name is shown in **bold type** above to create an initial **Main Topic** document on Saturday October 16 (the specific content is unimportant—a simple introduction will suffice) within their groupx.nsf database and add the Main Topic document to the server via a database exchange. **All** other group members will then add **Response** documents on Sunday via database exchanges (again, any message will do) to insure that their groupx.nsf database is ready for the preliminary analysis work. Howard Deckelbaum will be monitoring your progress during the weekend, and will provide assistance with any problems.

On **Monday October 18**, I will provide each group with their particular preliminary analysis responsibility. Members should use their own groupx.nsf database to discuss approaches, divide responsibilities, and formulate questions for Mike Mocciola or me. Both Mike and I will be actively monitoring (i.e., lurking) each groupx.nsf database to insure that work is progressing satisfactorily and that all questions for us are answered promptly. While each group deliberation is private, there will be questions or findings raised that should be shared with the workgroup as a whole, and these I will post in the virtual.nsf database.

The Virtual College
Course Session 3

Y52.1100 Systems Analysis
Alternatives Analysis of the Silicon Valley System

Analysis   Design   Development

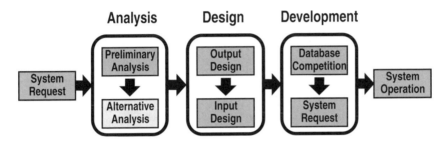

SESSION OBJECTIVES

The objective of this session is to propose and analyze a set of alternatives to the current Silicon Valley student data and communications system. When the findings of a preliminary analysis warrant a continuation of the systems study, an alternatives analysis is conducted. This analysis identifies the general scope and objectives of the new or modified system, presents the costs and benefits of two or more alternative systems, describes any organizational or policy changes of the proposed alternatives, and recommends one alternative for subsequent design and development.

As with preliminary analyses, alternatives analyses vary in scope and detail depending on organizational requirements. The alternatives analysis format that our workgroup will use is fairly generic, and will include the following sections:

1. **Introduction**
   Brief description of the purpose of the analysis and of the proposed alternative systems.
2. **Alternatives Analysis Recommendations**
   A summary of the alternative systems considered by the study, and the specific reasons (e.g., reduced costs, improved effectiveness) for recommending the implementation of one of them.
3. **Alternative Analysis**
   For **each** of the analyzed alternatives, a summary page using the **Alternatives Analysis Form** (click on the **Compose** menu and then on the

form name to see it) is prepared. This form will contain a description of the proposed alternative, a list of specific benefits and cost savings where appropriate, an itemization of the system's development and annual operating costs, and a basic cost-benefit analysis using the payback method.

SESSION ASSIGNMENTS (DUE NOVEMBER 2)

Readings: Hammer article ▯ (electronic)
Sproull and Kiesler, Chapters 4 - 6
Kendall and Kendall, Chapters 9 ▯ and 13 ▯ (electronic)
Case Study: Alternatives Analysis of the Database System ▯ (electronic)

Our goal during the next seven days is to identify and analyze **four** alternatives to the current Silicon Valley student database and communications system. In selecting these alternatives, we will not simply be improving the current system. Rather, we will attempt to reengineer the database system process in a manner similar to the process redesign efforts described in the Hammer article. We will need to challenge some of the underlying assumptions of the current system. Why it collects data or conducts meetings the way it does? Why certain people are involved or uninvolved in the overall process? Where are the main errors occurring and why?

In the interests of time, the workgroup will again be divided into four subgroups—Group A, Group B, Group C, and Group D—with each subgroup being responsible for preparing one alternative analysis. Workgroup members will remain in their original subgroups as shown in Table 1 below.

Table 1. Composition of Project Subgroups

GROUP A (groupa.nsf)	GROUP B (groupb.nsf)	GROUP C (groupc.nsf)	GROUP D (groupd.nsf)
Student 1	**Student 7**	**Student 12**	**Student 17**
Student 2	Student 8	Student 13	Student 18
Student 3	Student 9	Student 14	Student 19
Student 4	Student 10	Student 15	Student 20
Student 5	Student 11	Student 16	Student 21
Student 6			

Group members will have **Author** access to the group that they are listed under as shown in Table 1. As was done during the preliminary analysis phase of our project, members should discuss approaches, divide responsibilities, and formulate questions for Mike Mocciola or me. Both Mike and I will be actively monitoring each group's discussion to insure that work is progressing satisfactorily and that all questions for us are answered promptly.

By 11:00 pm on Tuesday, November 2 each group should submit to the designated location in the **virtual.nsf** database a completed **Alternatives Analysis Form** for their systems option. Please use the two Alternative Analysis forms included in the *Case Study: Alternatives Analysis of the Database System* report as guides for the preparation of your alternative. I will have more to say about the preparation of the **Alternatives Analysis Form** during the next few days.

The Virtual College
Course Session 4

Y52.1100 Systems Analysis
Output Design of the Silicon Valley System

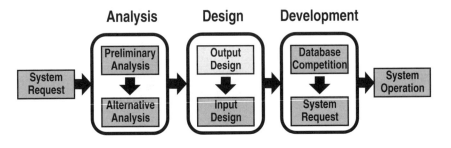

SESSION OBJECTIVES

The objective of this session is to design the initial set of outputs for the new Silicon Valley student data/communications system, based upon the information requirements of the center director. The output design stage of the systems development process determines the content and format of the various reports and files that the users need. Users provide ideas and analysts (us!) produce samples of output designs for the users' approval. Contrary to what some people think is logical, output precedes input. Further, output determines input. Users and analysts agree on the output (information) requirements of a system first, and then identify the inputs (data) necessary for the generation of the output.

In a Lotus Notes application, the **views** are the primary output, while the **forms** are the input (the data collection method). Although an analyst using Notes builds forms **before** views, he or she must know what the views need to contain before the forms are created, since all information in a Notes application is ultimately derived from data collected by and contained in the forms. Views are similar to database reports, and a database developer has to know what kind of reports an organization wants before the database is designed. Consequently, we will design **(in text)** the views first and then proceed to design and develop the forms **(in Notes)**. Finally, we will develop the views **(in Notes)** themselves.

Based in part on the capabilities that led to the selection of the Notes-based option in the alternatives analysis, there appear to be two initial outputs that need to be designed:

1. **Unit-Record Student Data File Formatted for Statistical Analysis**
   The corresponding Notes view will display the individual student records for export as a Lotus 1-2-3 (or similar) file. The view will display all data fields designated for statistical analysis.
2. **Discussion Database to Support Curriculum Development Workshops**
   The corresponding Notes view will display all documents created by participants in chronological (or other designated) order within various categories. The view will display such document characteristics as author, category, topic and date.

The output-input-output process described above will require the interdependent efforts of our four work groups. Each group will be responsible for designing one center system input or output. Groups A and C will assume responsibility for output design, and Groups B and D will be responsible for the input designs. The groups' products, formats and due dates are provided in Table 1 below.

Table 1. Group Output/Input Design Schedule

Group	Design Product	Format	Due Date
A	Student Data Record	Text Description	November 8
C	Discussion Database	Text Description	November 8
B	Student Data Input Screen	Notes Form	November 13
D	Discussion Input Screen(s)	Notes Form(s)	November 13
A	Student Data File	Notes View	November 16
C	Discussion Output Screen(s)	Notes View	November 16

As can be inferred from Table 1, the preliminary output design work of Groups A and C will affect the subsequent final input design work of Groups B and D, which in turn will affect the subsequent final output design work of Groups A and C. I will have more to say about this output/input design schedule in the coming days.

SESSION ASSIGNMENTS (DUE NOVEMBER 8)

Readings: Sproull and Kiesler, Chapter 7
Kendall and Kendall, Chapter 15 ☐ (electronic)

1. **Student Data/Communications Output Design.** Our goal during the next few days is to work with Mike Mocciola on determining the center's information (i.e., output) requirements for generating unit-record student data and for supporting the curriculum development workshops. Groups A and C will be responsible for this phase of the design effort, with Group A focusing on the student data elements that need to be generated as a computer file, and Group C focusing on the information requirements of the curriculum discussion databases. Both groups should work within the *Output Design* category area of their **groupx.nsf** databases, and they should make their questions to Mike Mocciola or me easy to find. I will be posting important outcomes of these group design sessions to the *Output Design* session area of the **virtual.nsf** database.

2. **Systems Project Request.** As indicated in the **Course Outline**, each student will design and build in Lotus Notes a working application system relevant to a job-related problem or need. To ensure that your intended application is appropriate both for Notes and for the assignment's scope and timeframe, I would like **each** of you to prepare an **Analysis Request Form**. Using the analysis request form prepared for the Silicon Valley database system as a guide ☐ , select from the **Compose** menu and complete a new **Analysis Request Form** for your intended application. Try to estimate as many operational parameters as possible. I will create within this *Output Design* session area a new public document titled **Systems Project Analysis Request Forms** to which you should link your Analysis Request Form (which functions as a response document). Please prepare and send the Analysis Request Form to the above document no later than **November 17.**

The Virtual College
Course Session 5
Y52.1100 Systems Analysis
Input Design of the Silicon Valley System

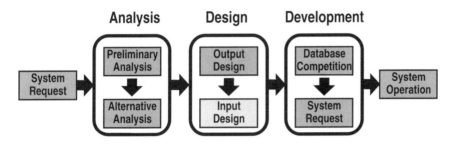

Analysis   Design   Development

SESSION OBJECTIVES

The objective of this session is to have groups B and D design the input forms for two Notes databases—Student Data and Workshop Topics—whose outputs were designed by groups A and C respectively In the interests of time, only the Workshop Topics application of the two databases proposed by group C will be developed as a functioning Notes database. Throughout this process (and the subsequent session on views design), I will describe the forms development procedures so that you can develop forms for your Systems Project application.

A summary of the groups' products, formats and due dates are provided in Table 1 below.

Table 1. Group Output/Input Design Schedule

Group	Design Product	Format	Due Date
A	*Student Data Record*	*Text Description*	*November 8*
C	*Discussion Database*	*Text Description*	*November 8*
B	Student Data Input Screen	Notes Form	November 13
D	Discussion Input Screen(s)	Notes Form(s)	November 13
A	Student Data File	Notes View	November 16
C	Discussion Output Screen(s)	Notes View	November 16

Completed activities are shown in *italics*.

SESSION ASSIGNMENTS (DUE NOVEMBER 14)

Readings: Sproull and Kiesler, Chapter 8
            Kendall and Kendall, Chapter 16 ⬚ (electronic)

The goal of groups B and D during the next five days is to work with their corresponding output design group (A or C) and Mike Mocciola on determining the center's input requirements for collecting unit-record student data, and for entering discussions in the Workshop Topics database. Groups B and D will be responsible for this phase of the design effort, with Group B developing the **single** Notes student data form, and Group D developing the **three** Notes workshop topics forms. Both groups should work within the *Input Design* category area of their **groupx.nsf** databases, and they should make their questions to Mike Mocciola or me easy to find. I will be posting important outcomes of these group design sessions to the *Input Design* session area of the **virtual.nsf** database.

   As a starting point, both groups B and D will need to get clarification about the output designs prepared by their counterpart groups A and C. I have posted the summaries of the A and C groups' output design efforts as **Response** documents to this **Course Session** document. Questions to and answers from the two group pairs (A - B and C - D) should be posted as **Response to Response** documents to the appropriate **Response** document. Mike and I will be adding our two cents' worth to these public **virtual.nsf** discussions as appropriate.

   All forms must be completed by **11:00 p.m. on Sunday, November 13**.

The Virtual College
Course Session 6

Y52.1100 Systems Analysis
Implementation of the Silicon Valley System

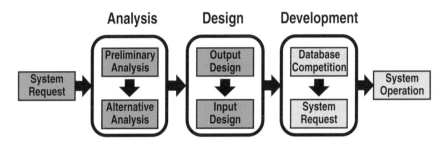

SESSION OBJECTIVES

The objective of this session is to complete the development of the Silicon Valley data/communications system. During the next seven days, we will produce the system's initial outputs (Views), and will create the databases' icon, policy and help documents. At the end of the week, the **Student Data** database will be tested at two project sites and the **Workgroup Topics** database will be reviewed at the Friday curriculum development meeting. During this week, I will describe the Notes views, policy and help development process so that you can develop these elements for your individual Systems Project application.

SESSION ASSIGNMENTS (DUE NOVEMBER 21)

Readings: Sproull and Kiesler, Chapter 9
Kendall and Kendall, Chapter 21 ▯ (electronic)

During the next four days, we need to develop two views—one view for the **Student Data** database that will export the individual student records as a Lotus 1-2-3 WK3 file, and one Main View for the **Workshop Topics** database. As these views will not have any of the data summaries that are typical of some Notes views, their development will be fairly straightforward. For the **Student Data** database, Mike Mocciola will provide some test student records, and he will import the exported Notes file into SPSS on his office PC to verify the file format.

Once the new system's forms and views have been completed (at least for purposes of the pilot test), we will move them from the **virtual.nsf** and **teamx.nsf** databases to a new database called Silicon Valley (a.k.a. **silicon.nsf**). You will then add this new database to your Notes workpage.

During the next few days, we must ensure that our database implementation efforts meet the following deadlines for the database pilot test.

PILOT TEST ACTIVITY	DUE DATE
1. Install Notes on Two Project PCs	November 14
2. Train Two Coordinators in Notes Data Entry Procedures	November 15
3. Install **Student Data** Database on Center PCs and Train Coordinators	November 16
4. Collect Sample Student Data at Project Sites	November 17
5. Enter and Edit Sample Data into Projects' **Student Data** Databases	November 18
6. Conduct Testfile.WK3 Database Exchange with NYU Server	November 19
7. Demonstrate **Workgroup Topics** Database at Curriculum Meeting	November 20

Acronyms

AAL	ATM adaptation layer
AAP	alternate access provider
ABR	available bit rate
ACD	automatic call distributor
ADSL	asymmetric digital subscriber line
AGSAT	Agricultural Satellite Corporation
ALIS	Ameritech LAN Interconnect Service
ANSI	American National Standards Institute
APC	Advanced Professional Certificate
API	application programming interface
APPN	advanced peer-to-peer networking
ARPA	Advanced Research Project
ASTN	Automotive Satellite Television Network
ASTS	Arts and Science Teleconferencing Service
ATM	asynchronous transfer mode
BARRNET	Bay Area Regional Research Network
BCSN	Black College Satellite Network
BISDN	broadband ISDN
BRI	basic rate interface
BT	burst tolerance
CAD	computer-aided design
CAE	computer-aided engineering
CALREN	California Research and Education Network
CAM	computer-aided manufacturing
CBR	constant bit rate
CCD	charged couple device
CD-ROM	compact disc–read only memory
CDV	compressed digital video

CDVT	cell delay variation tolerance
CERFnet	California Education and Research Federation Network
CIA	certified internal auditor
CICNet	Committee on Institutional Cooperation Network
CIDVL	Collaboration for Interactive Visual Distance Learning
CLEC	competitive LEC
CLID	calling-line identification
CMC	computer-mediated communications
CMD	circuit mode data
CNN	Cable News Network
CO	central office
CPE	customer premise equipment
CRS	cell relay service
CSCW	computer-supported cooperative work
CSNET	Computer Science Network
CSU	channel service unit
CTX	Centrex
DBS	direct broadcast satellite
DIAL	Distance Instruction for Adult Learners
DMT	discrete multitone transmission
DN	directory number
DNIS	dialed number identification service
DOD	Department of Defense
DOE	Department of Education (state-based or federal)
DOT	Department of Transportation
DQDB	distributed queue dual bus
DS1	digital signal 1
DS3	digital signal 3
DSU	data service unit
DVMRP	distance learning vector multicast routing protocol
EDP	electronic data processing
EFCI	explicit forward congestion indication
EISA	extended industry standard architecture
FCC	Federal Communication Commission
FDDI	fiber distributed data interface
FEP	front-end processor
FETN	Fire and Emergency Television Network
FRAD	frame relay access device
FRS	frame relay service

FSN	full-service network
FT1	fractional T1
FTP	file transfer protocol
FTTB	fiber to the building
FTTC	fiber to the curb
FTTH	fiber to the home
FTTN	fiber to the node
GUI	graphical user interface
HDTV	high-definition television
HFC	hybrid fiber/coax
HIPPI	high-performance parallel interface
HLI	high-speed LAN interconnection
HRM	human resource management
HSTN	Health and Sciences Television Network
IAD	integrated access device
IC	interexchange carrier
ID	identification
IDL	interactive distance learning
IDTN	Interactive Distance Training Network
IEEE 802.6	Institute of Electrical and Electronics Engineers 802.6
IETF	Internet Engineering Task Force
IIA	Institute of Internal Auditors
IMNG	Interactive Medical Network Group
I/O	input/output
IP	Internet protocol
IPS	integrated production system
IS	information systems
ISA	industry standard architecture
ISDN	integrated services digital network
ISO	International Organization for Standardization
ITU-T	International Telecommunication Union–Telecommunication
ITFS	instructional television fixed service
IXC	interexchange carrier
JPEG	Joint Photographic Expert Group
Kbps	kilobits per second
LAN	local area network
LAPD	link access protocol for ISDN D channels
LATA	local-access transport area
LEC	local exchange carrier

LETN	Law Enforcement Television Network
LTCN	Long Term Care Network
MAC	media access control
MAN	metropolitan area network
Mbps	megabits per second
MCET	Massachusetts Corporation for Educational Telecommunications
MCI	media control interfaces
MCR	minimum cell rate
MCU	multipoint control unit
MES	master Earth station
ME/U	Mind Extension University
MIME	multipurpose internet mail extension
MPEG	Motion Picture Expert Group
MSO	multiple systems operator
MTTR	mean time to repair
NCIH	North Carolina Information Highway
NEARNET	New England Academic and Research Network
NETO	National Educational Telecommunications Organization
NFPA	National Fire Protection Association
NFS	network file system
NI-1	National ISDN-1
NI-2	National ISDN-2
NI-3	National ISDN-3
NIC	network interface cards
NII	national information infrastructure
NIUF	North American ISDN User Forum
NMLIS	native-mode LAN interconnection service
NREN	National Education and Research Network
NRT	nonreal time
NSF	National Science Foundation
NSFNET	National Science Foundation Network
NTIA	National Telecommunication and Information Administration
NTSC	National Television Standards Committee
OLTP	online transaction processing
OS	operations systems
OSI	open system interconnection
OSIRM	Open System Interconnection Reference Model
PAD	packet assembler/dissassembler
PAL	phase alternate line

PBX	private branch exchange
PBS	Public Broadcasting Service
PC	personal computer
PCI	peripheral component interconnect
PCR	peak cell rate
PCS	personal communication services
PDA	personal digital assistant
PDU	protocol data unit
PIM	protocol-independent multicast
PIN	personal identification number
PMO	present mode of operation
POP	point of presence
POTS	plain old telephone service
PPP	point-to-point protocol
PRI	primary rate interface
PSTN	public switched telephone network
PVC	permanent virtual connection
QOS	quality of service
RAID	redundant array of inexpensive disks
RBOC	regional Bell operating company
R&D	research and development
RF	radio frequency
RGB	red, green, and blue
RSVP	ReSerVation protocol
RT	real time
SCR	sustained cell rate
SCSI	small-computer serial interface
SDH	synchronous digital hierarchy
SDS	switched digital service
SDTV	standard-definition TV
SECAM	sequential color and memory
SIP	SMDS interface protocol
SLIP	serial line internet protocol
SMDS	switched multimegabit digital service
SMTP	simple mail transfer protocol
SNA	system network architecture
SNI	subscriber network interface
SONET	synchronous sptical network
SPID	service profile identifier

ST-II	streaming II
SURANet	Southeastern Universities Association Network
SVC	switched virtual connection
SVHS	super VHS
TA	terminal adapter
TC	television center
TCP	transmission control protocol
TDM	time division multiplexing
TIIAP	Telecommunications Information Infrastructure Assistance Program
TLS	transparent LAN service
TSC	television switching center
TVRO	TV receive only
UBR	unspecified bit rate
UDP	user datagram protocol
UNI	user network interface
USDLA	United States Distance Learning Association
VAN	value-added network
VBR	variable bit rate
VDT	video dialtone
VIP	video information provider
VIU	video information user
VSAT	very small aperture terminal
VS/VD	virtual source/virtual destination
WAIS	wide area information services
WAN	wide area network
Westnet	Southwestern States Network
WHTG	Westcott Healthcare Teleconference Group
WWW	World Wide Web (server)
YUV	Y=luminance; U,V=color components used by PAL and NTSC

About the Author

Dan Minoli has extensive interdisciplinary experience in advanced telecommunications and data communications, which he acquired through tenures at a number of premiere technology-based organizations. His specialty is WAN and LAN broadband and ATM network design, development, engineering, and implementation.

For the past decade, Mr. Minoli has worked on many practical aspects of WAN and LAN networking. In addition to fundamental work in support of the development of ATM and frame relay, Mr. Minoli, director of engineering and development, has supported the actual deployment of Teleport's ATM WAN and LAN networks in 10 U.S. cities. Earlier, as a consultant at DVI Communications, he assisted several companies in deploying campus-based ATM networks. Still earlier, at Bellcore, he assisted a number of RBOCs in issuing and reviewing a series of RFPs and responses for local carrier ATM switches. He also has assisted RBOCs in performing architecture planning for public ATM and other networks to support digital video and high-speed data services. He also worked on the North Carolina information highway in support of distance-learning applications over ATM. In the 1980s, Mr. Minoli was heavily involved in DS0/DS1/DS3/router-based enterprise networks, particularly while at Prudential Securities, ITT Worldcom, and Bell Labs.

Mr. Minoli has written 21 books on enterprise networking, ATM, digital video, LANs, imaging, outsourcing, telecommuting, multimedia, and other topics. His books have enjoyed high industry circulation and critical acclaim and are used in over 20 universities.

Mr. Minoli has written several comprehensive market reports for Probe Research Corporation on topics such as frame relay and SMDS, multimedia, telecommuting, imaging, and ATM. He has also written the copy for several sales brochures for networking equipment companies.

Mr. Minoli is on Datapro's advisory board for broadband networking and over the last decade has published more than 50 *Technology Reports* on topics including ATM, frame relay, LANs, and video. He is also an advisor to DVI Communications and IMEDIA.

Mr. Minoli is a frequent speaker and session organizer at industry conferences and is quoted regularly in the press. He has spoken at more than 35 conferences on topics such as ATM and multimedia.

Mr. Minoli has written more than 220 technical and trade articles for numerous journals on topics such as enterprise networking, ATM, frame relay, and digital video. He estimates that one in five people in the industry has purchased one of his books, and that each individual in the industry has read at least one of his technical articles.

Mr. Minoli has been a contributing editor of *Network Computing* magazine for a number of years, as well as a columnist for *Network World* magazine, focusing on ATM and enterprise networking. He also was a columnist for *Computer World Magazine*.

On several occasions, Mr. Minoli has run a two-day seminar on enterprise networking; more recently, he has developed and delivered customized seminars for MCI, AT&T, Bellcore, and the RBOCs. In all, he has taught 30 professional seminars on communications, some of which were videotaped and have been widely distributed. He has held two nationwide, full-motion live-TV broadcasts on ATM technology, as well as conducted multisite videoconferencing corporate courses on enterprise networking. In all, he has done over 50 hours of instructional live TV.

In addition to full-time corporate affiliations, Mr. Minoli is an adjunct associate professor at New York University and has educated over 1,200 people on enterprise networking. He is also an adjunct professor at Stevens Institute of Technology, has taught courses at the Rutgers Center for Management Development, and has lectured at Carneige-Mellon University, Monmouth College, the University of Utah, and the University of Colorado. In all, he has taught 35 college courses.

Index

The Artech House Telecommunications Library

Vinton G. Cerf, Series Editor

Writing Disaster Recovery Plans for Telecommunications Networks and LANs,
   Leo A. Wrobel

X Window System User's Guide, Uday O. Pabrai

For further information on these and other Artech House titles, contact:

Artech House
685 Canton Street
Norwood, MA 02062
617-769-9750
Fax: 617-769-6334
Telex: 951-659
e-mail: artech@artech-house.com

Artech House
Portland House, Stag Place
London SW1E 5XA England
+44 (0) 171-973-8077
Fax: +44 (0) 171-630-0166
Telex: 951-659
e-mail: artech-uk@artech-house.com